M01 40002 07622

DATE DUE			
OCT 9 1985			
MAR 1 - 1989			
OCT 29 1990			
FEB 24 1993			

DEMCO 38-297

GOLD RUSH DAYS WITH MARK TWAIN

GOLD RUSH DAYS

WITH

MARK TWAIN

BY
WILLIAM R. GILLIS

With an Introduction by
CYRIL CLEMENS

Illustrated with woodcuts by
H. GLINTENKAMP

NEW YORK :: MCMXXX
ALBERT & CHARLES BONI

Republished, 1970
Scholarly Press, 22929 Industrial Drive East
St. Clair Shores, Michigan 48080

Library of Congress Catalog Card Number: 77-131719
Standard Book Number 403-00606-6

COPYRIGHT, 1930, BY
ALBERT & CHARLES BONI, INC.

Printed by *the Van Rees Press in the U. S. A.*

This edition is printed on a high-quality,
acid-free paper that meets specification
requirements for fine book paper referred
to as "300-year" paper

CONTENTS

Introduction	xi
Experience of a Pocket Miner	1
Steve Gillis — Printer	17
San Francisco in the 50's	22
Vigilantes Organized	29
Steve Goes to Oregon	35
When Chief Meets Chief	42
Enter Samuel Clemens	48
The Fight That Made Mark Twain Famous	53
Mark Goes to Jackass Hill	59
Mark Twain and the Burning Fuse	62
Clemens' One Mining Venture	65
Fun on the Hill	68
Conditions Along the Comstock	78
Lee Invites His Old Schoolmate to Visit Him at Virginia City	84
Rivalry Between Express Companies	92
Law and Liquor	94

Mark Twain's Boxing Bout	97
Dan McQuill Follows Mark's Nose	102
Mark Throws Up His Hands	108
I Go into a Funk	117
Mark Lectures on the "Holy Land"	120
Spiritualistic Manifestations	123
Mark Twain's Fake Duel With Sam Leonard	135
I Put a Hot Pitch Plaster on Doctor's Back	144
Mark Twain Becomes Private Secretary to Senator Stewart	148
Dan and I Witness the Piute Card Game	157
"Uncle" Jimmy Fair's Peculiarities	162
Mark Twain and the Goat	170
Mark Twain at a Wedding	175
Truthful James Spins a Yarn	178
The Entomologist and the Yellow-Jackets	211
Who Stole the Ham?	223
They Broke the Jug and Quit	236
A Sudden House Moving	242
How Uncle Bob's Prayer Brought the Rain	246

Introduction

THE old-timers who knew Mark Twain in the early days are not many. Just lately Becky Thatcher (Mrs. Laura Frazer) died, and still more lately there passed away Norval, "Gull," Brady. The following memoirs were written only a few months ago by William R. Gillis who was a friend and mining partner of Twain in the early Californian mining days. Gillis first met Twain when the latter went out to visit Angels' Camp in the winter of 1864. It was sitting around the tavern bar at Angels' that Mark Twain one winter night heard the story of the Jumping Frog. Mr. Gillis lived all his long life — he numbers ninety summers — just outside of Angels' at a village called Tuttletown. He is the custodian of the old cabin on Jackass Hill where Mark lived while a miner. Gillis lives a few miles away from the Mark Twain cabin, in a charming spot surrounded on every side by fair hills.

The serene peace and quiet of those hills have taught Gillis much wisdom and no little philosophy. Every time I speak with him I catch myself thinking of him as holding the chair of philosophy in one of our large universities.

The passing years have not dimmed the memory of Mr. Gillis. When he speaks of his friendship with Mark Twain we can rely upon what he tells us. It is fortunate for all lovers of Mark Twain that when those who knew him in his early mining days in the West have almost left us, we have these crisp, fresh, fascinating accurate recollections of one who knew him as a fellow-miner in those romantic days of the after-glow of the great California Gold Rush.

As a sort of preface to Mr. Gillis' own account I quote what Mark Twain himself says about Angels' Camp in "Roughing It":

"By and by, an old friend of mine, a miner (this was Jim Gillis, an elder brother of William) came down from one of the decayed mining camps [Angels' Camp, is meant] of Tuolumne, California, and I went back with him. We lived in a small cabin on a verdant hillside [Jackass Hill], and there were not five other cabins in view over the wide expanse of forest and hill."

A little later Twain goes on to say:

"At last we shouldered our pans and shovels and struck out over the hills to try new localities. We prospected around Angels' Camp, in Calaveras County, during three weeks, but had no success. Then we wandered on foot through the mountains . . . . but still we remained as centless as the last rose of summer."

A few months ago Mr. Gillis lost his wife, his faithful companion for sixty years. Mr. Gillis told me that whenever a stranger knocked at her door, his wife did not give him a mere handout, but insisted that he come in and sit at their table to partake of the best they had. I quote from "Roughing It":

"In accordance with the custom of the country, our door had always stood open and our board always welcome to tramping miners — they drifted along nearly every day, dumped their paust shovels by the threshold, and took 'pot luck' with us — and now on our own tramp we never found cold hospitality." Book Two, Ch. 20.

Altogether this was a glorious period in the history of the West, and in this book of Gillis we are given a fascinating picture of a typical group, especially interesting because it numbered among its members the great Mark Twain.

CYRIL CLEMENS

Mark Twain Society,
Webster Groves, Mo.
June 23, 1929.

GOLD RUSH DAYS WITH MARK TWAIN

Experience of a Pocket Miner

LURED by "Gold, gold, hard and cold" I left San Francisco on May 4, 1863 to try my luck as a pocket miner in Tuolumne County. Taking passage on the old steamer *Paul Pry* I landed at Stockton the next morning and at six o'clock of the same day boarded one of the six-horse coaches of the old Sisson Stage Company and started on the sixty mile ride to Sonora. I shall never forget that ride as long as I live. Bumping over rocks, into chuckholes, through blinding clouds of dust and sweltering heat.

I was sitting on the seat with the driver, "Missouri Bill." We had not more than fairly started when he turned to me with, "I say, young feller, is you totin' a gun?"

"No," said I, "I am not, why do you ask?"

"Oh, nothin' much, ony I wus jest goin' to tell you, if you was you'd better drap it down by your feet or else git inside the coach."

"Why should I do either? Are you afraid the gun will go off and kill you?"

"Now, look here, boy, I wanna tell you, 'at I ain't afeard o' nothin', and I wasn't thinkin' 'bout your doggone gun goin' off at all, but wus jest s'posin' if we got hung up by stage robbers, 'tween here an' Sonory, you might be fool enough to pull it and so git yourself, an' me too, pumped full o' lead, that's all."

"Have you any reason for thinking that the stage may be held up by robbers this trip?"

"Yes I has, an' a mighty good one. Every time the Express Company makes a big shipment o' money, these fellows some way o'ruther knows all about it an' they knows jest as well as the Company does, that they's more'an fifty thousand dollars in the two boxes on the stage to-day an' it's a two to one bet they'll try to get it so you just keep your eye peeled, son, an' ef you see a man step out from a clump o' bushes or from behind a rock, throw up your hands as high as you kin, an' slide down on the foot board and lay low.'"

"While I am throwing up my hands and lying low, what will happen to you?"

"Say, do you see that hoss fly on my off leader's ear? You just watch him." With these words the lash of Bill's whip shot out and flicked off the fly without touching the horse.

"Now," said Bill, "that's what I'll do to any son of a gun who tries to hold up this stage. Before he can get his gun on me, my lash will take him between the eyes; next it will go to the backs of my leaders, then — 'good-by John,' to the robber, and the treasure is saved."

Without the experience of a "holdup" or witnessing Bill's dexterity with his whip on a robber, I arrived in Sonora at six o'clock that evening tired, hungry and dirtier than ever before in my life.

The next morning, after providing myself with a kit of tools comprising a pick, shovel, crowbar, sledge hammer and gold pan, accompanied by my brother Jim, I took the trail to "Jackass Hill" to take up the life of a pocket miner.

The afternoon of that first day on the Hill, I spent in getting acquainted with the people living there and my reception was so kind, so friendly that I felt, as their warm hand-clasps responded to mine, and their hearty voices welcomed me to the Hills, I was among a company of old friends instead of strangers.

I moved into the cabin with Dick Stoker that evening. The next morning, while Dick and I were at breakfast, Jim burst into the cabin with, "Billy, if you expect to make anything pocket-mining, you've got to get out earlier than this time o' day; loafing around ain't going to dig out any gold. Shoulder your tools and come on and I will show you a place to work." And turning to Stoker with a grin, said, "I'm going to break him in on the Van Houghten, Dick."

"Well, Mr. Gillis," said Dick to me, "there is lots of quartz on the Van Houghten and as gold pockets are found in quartz, *perhaps* you may find one there and I hope you will."

With these parting words from my cabin mate, Jim and I sallied forth for the place of my future mining operations. Arriving on the ground, Jim said to me: "Now, Billy, here's a big, bold vein of quartz; there's room enough in it for millions of dollars worth of gold, if you are lucky enough to find it. There is one thing I want to tell you though, before you begin work. You will know gold when you see it; if you find anything which you think is gold, but of which you are in doubt, throw it away. Now, take your pick and start in by digging out those blocks of quartz. As you dig them out, break them with your sledge into pieces about the size of a walnut; while doing this keep a sharp lookout for the gold that may be in them. Put some of the broken rock in your pan, and wash it in the ditch once in a while. If you fail to get a color in the first pan, don't let it discourage you; come back and get another pan."

By noon I had pounded about half a ton of rock into bits, and washed about a half dozen pans in the ditch, and all I had to show for my work was sore muscles, blistered hands and my fingers all cut with small fragments of sharp quartz. When I reached the cabin covered with dirt and reeking with sweat, Dick greeted me with:

"What luck?"

"If you call this luck," I answered, "I've had lots of it," showing him my blistered hands and bleeding fingers. Then I told him of my half day's strenuous labor.

"You certainly have got your hands into a pretty bad fix by pounding and panning that quartz and I am afraid that you'll have a lot of trouble with them. Go over to the sink and wash the dirt out of the cuts and I will see what I can do to relieve the pain and neutralize the poison in them."

"Poison, why how can there be poison in them, Mr. Stoker, and if there is, where did it come from?"

"From the quartz, of course; there is more or less poison in all quartz, and a cut from it will sometimes cause a bad sore, unless it is promptly attended to."

Dick then went to his trunk and took therefrom a small tin box containing some kind of sticky substance which he spread upon strips of cloth, and carefully bound up my sore fingers.

"Now you had better knock off for the rest of the day," said he, "and rest up."

I felt very much inclined to take Dick's advice and rest up until the next morning, but after eating a hearty dinner, and drinking two cups of strong coffee, I started for my diggings again.

"Hello," said Dick, "going back to work?"

"Yes," I replied, "I will try it, for awhile, anyhow. I've got a hunch that I'm going to strike that pocket, so don't fall over if I bring a thousand or so back with me."

"Ha! ha!" laughed Dick. "Good luck to you. Here, put on these gloves, they will save your hands; and don't break them rocks to pieces. Just come down on them hard enough to crack them, if there is any worthwhile gold in them they will break to it."

When I got to the claim I took up my hammer and began breaking the rocks according to Dick's instructions. I had been hammering away on them for perhaps an hour, without results, when I pried out an extra large one and coming down on it with a mighty whack it came apart with a sort of whine and there it lay before me held together by strands of coarse gold. To say that I was excited when my eyes beheld that golden vision — Well, I just dropped my hammer and, shouldering the big boulder, started at a five mile clip for the cabin.

When I burst in at the door, I startled Dick out of a nap by yelling at him, "Hey, partner, how's this for a prospect?" When Dick's eyes caught the gleam of the yellow metal his excitement nearly equaled my own.

"Prospect! Great Scott, man, that's not a prospect, it's *gold* and big gold at that. Why, you've struck it rich. Say, when you get another hunch let me in on it, will you? Talk about luck, why, I never heard of luck like this. A pocket on your first day's work in the mines!"

My "clean-up" from this one rock amounted to a little over seven hundred dollars. I picked and shoveled, broke rock and panned, for three months after this "lucky strike" but my "hunches" all went back on me, and my eyes were never again gladdened by seeing another color of gold. I finally abandoned the Van Houghten and moved to another locality on the Hill, and, at Jim's suggestion, went to tracing for gold on the surface.

"Billy," said he, "begin panning at the foot of

the hill; if you get a color it is a prospect and indicates that there is gold above. Take a pan higher up and so continue working up hill until you get above your prospect, then go back to where you got your last one and dig down to bed rock, then trench along uphill until you find the vein."

"And when I find the vein, what then?" said I.

"Why, pan along it, of course, until you reach the point where your prospect begins to increase, then be careful and watch out for your pocket."

Jim then left me and went to Sonora. At about three o'clock that afternoon, I struck a small quartz vein. I took a pan and went to my pan hole and washed the sand and dirt out and there, in the bottom, I saw three nice little specks of gold. Jim had told me that a color was a prospect. "Now," I reasoned to myself, "if a color is a prospect, three colors must be pretty close to a pocket and, as I might throw away a lot of gold, I had better not dig any more, but wait until Jim gets back to help me take it out." And I went home and placed the pan under my bed, and began "building castles."

It was long after dark when Jim walked into the living room at Carrington's, but I was there waiting for him. My head was so full of the gold in the pan hid under my bed that I did not even wait for him to be seated before I burst out with: "Jim, I've got a prospect down at the cabin that — when you see it — will make you think your brother Billy is no slouch as a pocket miner. It's a corker, you bet, and I want you to be on hand early in the morning to help me dig it out." The next morning, Jim was at the cabin before sunrise, and found both Stoker and myself still in bed.

"Here, Billy," he shouted, "get out of that bed and show me your *corker* of a prospect." I reached under the bed for the pan and handing it to him, said: "There, feast your greedy eyes on gold, Jim, while I make the fire."

Jim took the pan and, going to the door where there was better light, scanned the bottom for my big prospect, when he saw my three nuggets, disgustedly snapped: "Well, I'll be darned — *three flyspecks.*" And, without another word, dropped the pan and marched away. When he said, "Three flyspecks" and went off in the contemptuous way he did, Jim let himself out with me either as an expert or instructor in pocket-mining, and I determined to depend upon my own judgment in the future and, by close observation, I soon became familiar with the character of gold-bearing quartz veins and formations in which they were incased, together with their crossings and contacts. What I learned, however, availed me very little in finding gold. With the exception of one or two small "bunches" my income was nil.

In August, 1863, Carrington, Jim, Stoker and I entered into a partnership which Stoker called "The Jackass Hill Syndikite of Mine Workers." That name should have by all means brought us good luck, but it didn't.

After prospecting all over Jackass Hill and the surrounding country until November, our Company went broke. All we had left to show for our time and labor was an unpaid bill at the store.

Upon the dissolution of the "Syndikite" Jim went to San Francisco and Dick resumed work on the "Stoker Mine," leaving Carrington and me to carry on as best we could. We then decided to try our luck in Calaveras County and, shouldering our tools, we hiked over to Angels' Camp, making the Union Hotel our headquarters.

During the next month we prospected along the Mother Lode until December 24th when, having met with no success, we returned to Jackass Hill and resumed prospecting there. But, until the 8th day of January, 1864, our *bad luck* still held *good*. On that day we found a prospect that developed into a pocket from which, in the next three days, we panned out seven thousand dollars. The gold from this pocket relieved us from all financial trouble and made me regard the world a pretty good old world after all.

From this time until August, 1867, when I left Tuolumne County for Virginia City, I continued to prospect for quartz pockets with varying success, sometimes sailing before a fair wind over a sea of prosperity, at others wallowing, rudderless, among the breakers of adversity.

At this time Tuttletown was one of the liveliest towns in the county, mining, both quartz and placer, being prosecuted vigorously with paying results to the miners. The ten stamp mill on the Patterson Mine was steadily running on good ore, and the placer-miners in the ravines and gulches were reaping good harvests of gold from their claims and an air of prosperity hung over all. There was one store, carrying a full stock of merchandise such as groceries, boots and shoes, clothing and mining supplies; a hotel and a barber shop, and of course, a saloon, where wine and whisky were dispensed to the thirsty populace. Besides these business houses, there was a Literary Society, with a membership of three hundred, having a library of near a thousand volumes of standard prose and poetical works.

Under these prosperous conditions there was, of course, a full quota of "Soldiers of Fortune," otherwise known as "Poker Sharps" gathered to help the "Greenhorns" get their money into circulation.

One of these sharps was a suave, handsome chap named Tom Lucket. The saloon was owned by Valentine Pitorf, who afterwards opened and ran Turnverein Hall in Sonora. Now Valentine, in his own saloon at Tuttletown, was a sober, steady business man, never indulging to excess in his liquid refreshments, but once in every month his thirst would overcome him and, leaving his business in Lucket's care, he would go to Sonora to quench it — which was usually a three days' job.

On one of these occasions, as Valentine was about to mount his horse, he said: "Tom, while I am gone you run the saloon just like it was your own."

"All right, Val, I'll do the best I can," replied Tom. Valentine had been gone about two hours when Jim Moran, old Pat's amalgamator, entered the saloon and asked all hands to take a drink. This man, Moran, had been working in the mill a year or more at this time and had gained the goodwill of every one by his good nature and friendliness, taking part in all social gatherings and sports; he was a regular attendant at all the meetings of the Literary Society, joining in our debates and other proceedings, willing to sing a song, read a poem, dance a jig, or do anything else that would contribute to the pleasure of the evening. His visits to the saloon were very infrequent; on the occasion of these visits he would sometimes play a game of billiards or join in a game of pool but would have nothing to do with cards, saying they were "mighty poor tools for a working man to handle."

After drinking with the boys on this last occasion, he took a seat near a table where Tom Lucket, Frank Gross, and John Banks were playing a three-handed game of poker. As Moran seated himself Gross hailed him with, "Come, Jim, be a sport, get in and make the game four-handed."

"No, Frank, I guess I won't. I know nothin' about the game and would only make an ass of myself and be a joy killer to you fellows, so I think I will just watch your game awhile and go home."

"Aw, git in, Jim, an' play a few hands anyhow, just to show you got nothin' against us," said Banks. " 'Tain't a-goin' to hurt you any."

"I am not afraid of gittin' hurt, boys, but I'm — oh, well, I will take a hand in your game this one time but, remember, it will be the last time. I warn you also that at any time I see fit to quit the game I'm going to do so." With these words, Jim drew his chair up to the table and the game began.

The game had been going for some hours with Moran steadily losing and apparently becoming more and more excited every moment. Finally, at about 3 o'clock A. M., he appeared to lose his head entirely, betting his money so recklessly that the three sports themselves regarded it as almost robbery to rake it in.

At last Jim went ten dollars blind. Lucket straddled it; the other two stayed out and Jim made good. The cards were dealt and Lucket passed. Jim bet fifty dollars. Lucket saw the fifty and called for two cards, Moran taking three. After the draw, Tom bet twenty dollars.

"Twenty all you betting, Tom?" asked Moran, a sneering smile lighting his face.

"Don't bother your head about *my* bet, Jim, attend to your own."

"All right, Tom. I see your twenty and go you a hundred dollars better." Tom raised him back another hundred, taking the money from the till for the raise, his own being exhausted by covering Moran's bet. When Tom made this last raise, Moran picked up his hand from the table, leaned back in his chair and for fully five minutes intently studied it.

"Well, what you going to do?" at last snapped Tom.

"Well, it's either make a spoon or spile a horn. There's your hundred, Tom, and five hundred better. Play cards!"

"I ain't got the money to call that bet, Moran, and you know it, and you are trying to bluff me out, but it don't go. I'm sticking up the saloon and all that's in it against your five hundred and calling your bluff. What you got?" said Tom, throwing his hand on the table, exposing three queens and a pair of tens.

"A full house, eh, Tom?" said Moran. "But I guess these four little eights gets away with the cheese," and, raking the pot into his hat, he rose from the table and went behind the bar. After helping himself to a cigar he turned and said to the crowd: "Boys, I don't want the stuff in this shebang, so I'll just turn it over to you fellows. Drink it up or throw it out, just as you please. I am going to my cabin and turn in." But Jim did no such thing, he just kept on down the road and Tuttletown knew him no more.

When the three sports found that, instead of a

foolish fly, they had caught the biggest kind of a bald hornet in their net and had been badly stung him, they, too, left the saloon. Then began the wildest orgy of drinking and carousing ever known in Tuttletown and continued until Valentine arrived at about three o'clock, with murder in his eye, and cleared the room with a billiard cue.

Steve Gillis — Printer

My father and his elder children including myself landed in San Francisco in 1849. My mother and the younger children, one of which was Steve, then fifteen years old, landed April 22, 1853.

In July Steve went to work as a printer on the *Herald,* owned and published by John S. Nugent.

In 1853 San Francisco was the most lawless city in the United States. Criminals of every degree, from the "holdup" man to the petty thief, gathered there. Nearly the whole of Happy Valley was populated by ex-convicts from Australia, known as the "Sydney Ducks." These men and women were of a type bestial, bodily and mentally, unclean. Their children were leprous offspring of leprous parents. Growing up in an environment of debauchery and crime, they became the notorious gang of thugs and thieves known as the "Tar Flat Hoodlums," whose depredations were widespread over the territory south of Market Street, and were a terror to the decent residents of that locality. Belated wayfarers were waylaid and robbed, beaten into insensibility and left where they fell. Women and girls were grossly insulted on the streets, and on two occasions when an officer interfered he was assaulted and cruelly beaten.

These were the conditions when a young giant named McGowan was placed on duty south of Market Street. I will give my readers, as near as I can remember, his experiences with these "choice spirits," told in his own words in an interview had with him by my brother Jim and myself at Sonora.

"When I reported for duty after my appointment I was told by Sergeant Ayres that I was wanted in the chief's office. When I entered the room and announced my name, the chief looked up from the papers he had been examining and without a word, began looking me over. After about five minutes spent in summing me up he told me to be seated. Then he began to question me, asking me about everything I had ever been or done from the time of my grandfather down to the present. After this he instructed me as to my duties, and told me what he and the city expected of me as an officer.

"He then said, 'Mr. McGowan, you look to me like a pretty husky young fellow, and should be able to hold your own in any emergency. From the account you have given me of yourself, I think that in time you will become an efficient member of the police force of San Francisco. There are one or two localities in the city infested with a mighty tough lot of citizens. Brave, strong and active men are badly needed there, but as you have, as yet, had no experience as an officer, I will, for the time being, have you assigned to a nice, quiet neighborhood, where your duties will not be so strenuous while you are gaining experience.'

"He then gave me his hand and dismissed me, telling me to report for duty that night. By the smile on his face and the twinkling of his eyes while he was talking to me about that 'nice, quiet neighborhood,' I made up my mind that he had the 'joker' up his sleeve and was putting something over on me.

"The next night when I reported at the captain's office he introduced me to Officer Billy Blitz, and told me that he would serve in the same locality. My beat was from Market to Mission, and Billy's from Mission to Brannan, between First and Fifth streets. For the first three nights I was there I had no trouble whatever. On the fourth night, however, as I was walking along Howard Street, I noticed six 'boys' lined up across the sidewalk just ahead of me.

"As they showed no disposition to move when I

approached them, I said to them, 'Boys, you are obstructing the sidewalk. Stand aside and make room for other people to pass.'

"Then they began to get ugly, and called me some mighty rough names. I saw that they meant trouble but I kept a grip upon myself, answering nothing back, watching for their next move. It was not long coming. Finally, a big red-headed lad, who appeared to be the leader, said to me, 'We are not taking any orders from you, you big stiff,' and then made a grab for my helmet, but his jaw came up against my fist with such force that he went to the ground in a heap. Two of the others then came at me at the same time, but the toe of my boot took one of them under the chin, and I smashed the other with my right, and they both joined the red-headed fellow on the ground. I then felt another one of them on my back fingering for my throat, so I just reached over my shoulder, and grabbing him by the scruff of the neck, drew him in front of me and shook him till the breath was about all out of him. I then gave him a punch in the stomach and piled him up with the others. When I looked around for the two other lads I found they had crossed to the opposite side of the street.

"After I had the young rascals where I wanted them, I blew my whistle for Billy Blitz and he came to me on the run. In a short time after Billy joined me the four of them were riding to the station on the prison van."

"Mac," I asked, "did this scrimmage end your troubles with that gang of toughs?"

"No, sir, it did not but it was a good starter. For nearly two years after that Billy and I had our hands full with them, but when we were transferred to another part of the city those lads were as civil and well-behaved as any lot of Sunday school kids."

San Francisco in the 50's

ANOTHER hotbed of iniquity was the "Barbary Coast." Here were nightly assembled the vilest gang of criminals the world over, men and women, devoid of humanity and pity, robbing and killing for what they found on the persons of their victims. Woe to the stranger with anything of value in his possession who entered one of their dens alone! Trapped like a rat, he would be robbed, murdered, and his body thrown into the bay. There was seldom a week passed without a dead man being found floating in the bay, while a crushed skull, knife wound, or a cord twisted around his neck plainly told the manner of his death.

I am sure that such was the fate that a friend and former partner of my own met as late as October, 1868. He had come to San Francisco from one of the southern counties, where he had been getting out railroad ties for the railroad. He had in his possession $600, which he had earned at this work, and was on his way back to Tuolumne County, where he intended to resume work as a pocket miner.

I met him the day of his arrival in the city, and in the evening went to the theatre with him. After the performance we walked together from Washington to the corner of Montgomery and Clay streets. Here we parted, I going to my room on the corner of Powell and Clay streets, and he, as I then thought, to the "Russ House," where he had secured a room.

Before separating we agreed to meet at the "Russ" the next morning at 10 o'clock. On entering the office at the hour agreed upon my friend was not there. On going to his room and finding that it had not been occupied during the previous night, and knowing that he had all his earnings in a belt buckled round his waist, I at once surmised foul play. Telling Brush Hardenburg of my apprehensions regarding my friend, he accompanied me to the police station, where we reported the case, and asked the police to look up the missing man. They promised to do so, but after combing the city for a week reported that they had found no trace of him. He was never seen or heard of again.

There is no doubt that after parting from me on that fateful night he fell in with one of the Barbary Coast demons, who lured him into one of the death traps, where he was robbed, murdered and thrown into the bay. His body was probably carried out on the tide through the Golden Gate into the Pacific Ocean.

Gambling was run wide open throughout the city. Every game of chance known in the world was played there, from faro down to craps. There were more than a score of gambling hells in the city where the "hogging" game of faro was played.

Principal among these was "El Dorado," located at Kearney Street between Washington and Clay. This place was packed day and night by men of every vocation; working men, business men and men of leisure gathered there by the hundreds, all trying to harvest a portion of the great wealth displayed on the tables. There were twelve of these tables, every one of them having its full quota of the votaries of Dame Chance.

Placer mining in California was then at its apex and the production of gold from this source was enormous.

Thousands and thousands of dollars worth of this metal found its way to San Francisco in its virgin state, brought there by miners who had washed it from the earth. Some of these miners came to spend their money in having a "good time"; others were on their way to their families and sweethearts in the East with the earnings of years of toil and privation in the mines.

Most of the gold found its way into the coffers of those who ran these games of faro. At each table the dealer had a pair of gold scales beside him, and when a miner would place his bet on a card, the gold would be weighed, and in case the miner won the dealer would pay, placing a valuation of sixteen dollars per ounce on the gold, none of which was worth less than eighteen dollars, some of it as much as twenty dollars, so, whether the miner won or lost, the gambler got at least two dollars of his money.

All of these hells had men known as cappers, employed to keep a lookout for these men from the mines, and upon their arrival, to lure them to the games. Every boarding house, all the saloons and dance halls, in fact every place of public resort in the city, had a representative from one of these "tiger's dens." A capper would at once "spot" a miner among a crowd in any of these places, and if he found that he was not with a friend or party of friends, would approach and scrape an acquaintance with him, sometimes in a friendly way, calling his attention to something taking place in the crowd around them, then asking him to drink.

He would finally propose a walk to take in the sights, claiming that he, too, was a stranger in the city, and did not care to make the rounds alone at night, saying that it was not safe to do so. The miner, pleased to have a friend with him, would at once consent. After an hour or so passed in sight-seeing, the stroll would wind up at a faro game. Here the man from the mountains would soon play in all his money. Broke, and filled with booze, he would stagger into the street, where he would probably be arrested as a common drunk and pass the rest of the night in the city prison.

Prominent gambling houses were those of Steve Whipple on Sacramento Street, between Kearney and Montgomery; Billy Burrows' on Montgomery, north of Bush; and that of Colonel Jack Gamble on Bush Street, between Montgomery and Sansome. These games were ostensibly run on the square, and were supposed to be patronized by gentlemen only. But these "square" games always got away with the gentlemen's money, and were the cause of several of them serving terms in the state's prison at San Quentin.

Gambling was bad enough, but it was not the heaviest burden under which the people of San Francisco were groaning. Crime and gross immorality held carnival. Men were held up and robbed in broad daylight on some of the principal streets. Women of the underworld brazenly promenaded the streets and openly solicited for the houses of ill fame. One of these women, the famous Rose Cooper, employed a brass band to give two concerts a week from the balcony of her mansion on Pike Street, when the "girls" would go into the assembled crowd and invite the "boys" to enter.

The water front was lined with low sailor boarding houses into which the sea-faring man was drawn. When his purse became empty — the most of his money having gone into the till of his landlord — he would be drugged and shipped to sea again, the landlord drawing most of his pay for the voyage, claiming the sailor was in debt to him for the amount drawn.

The theaters, with the exception of one or two, were pandering to the lowest instinct of the human race, two of them, the Lyceum and Bella Union, nightly staging plays of the vilest and most indecent character.

There were scores of places where the brutal sports of rat killing and dog and chicken fighting took place. For the conditions existing in San Francisco, the community had no remedy. The courts were powerless to punish law breakers who had sufficient means to put up a defense. The halls of justice swarmed with professional jurymen, and a rich criminal had no difficulty in securing as many of these on his jury as were needed to block a conviction.

All this culminated in the assassination of James King of William by James P. Casey, an ex-convict from New York, who was then publishing a newspaper in the interests of the lawless class in San Francisco.

Mr. King had been exposing Casey's criminal life in the East in the *Evening Bulletin*. Casey went to the office of the *Bulletin* and demanded that King quit persecuting him, saying that if he did not stop publishing his scurrilous articles he (Casey) would kill him. Mr. King published the interview with Casey and kept right on exposing his past. Casey, exasperated by Mr. King's course, on the 14th of May, 1856, put his threat into execution by shooting him down in cold blood as he was walking along Montgomery Street near Washington.

Vigilantes Organized

WHEN it was made known that Mr. King had been murdered by Casey, the people of San Francisco went wild with anger and grief. Within forty-eight hours after the murder, at the call of "33 Secretary," more than five thousand of San Francisco's best citizens met and organized a "Committee of Vigilance."

When Casey was taken to the county jail, Charles Cora, confined there awaiting his second trial for the murder of General Richardson in the Blue Wing Saloon the previous year, became crazed with fear, imploring the sheriff to take him out of the city, saying, "Dave! For God's sake, get me out of here. Unless you do, my life is not worth a white chip, and Jim Casey will be my murderer as well as King's. Give me a chance, Dave! Give me a chance!"

Sheriff Scannel did everything possible to reassure him, but to no avail. He refused to be comforted. He at last quieted down, but until he was taken from the jail, maintained a gloomy silence.

This man Cora was the paramour of the notorious courtesan, Belle Cora. Upon his first trial the jury had failed to agree upon a verdict. He was remanded to the county jail to await a second trial, and although confined without bail, his treatment was more like that of an honored guest than that of a cold-blooded murderer. His sleeping quarters were all that a gentleman could wish, his table was supplied with every delicacy, the finest wines, liquors and cigars were on his sideboard. Here, of an evening, his friends gathered and the time was passed with music, drinking and poker playing.

The Vigilance Committee wasted no time in speechmaking, but immediately went to work on their great task. After raiding the armories of the National Guard and securing such weapons as were needed, they proceeded to the county jail and took both Casey and Cora, transferring them to "Fort Vigilant" on Sacramento Street, where they were held in close confinement until their trial and execution on the 22nd day of May, 1859, just eight days after the wanton murder of Mr. King.

While thousands of sorrowing citizens were following the remains of Mr. King to his last resting place in Lone Mountain, other thousands were gathered in the vicinity of Fort Vigilant to witness the execution of his slayer, together with that of Cora. While these two men stood upon the gallows trap, strains of solemn music came floating through the air, telling them that Mr. King was journeying to his long home. As long as the music could be heard the crowd of assembled thousands stood with uncovered heads. When it ceased the trap was sprung — the two men fell to their death. A solemn "Amen" of approval went up from the many gathered to witness the end of the great drama.

This was the beginning of San Francisco's regeneration. After hanging two more men — Hetherington, for the murder of Dr. Randall, and Brace, the slayer of Marion — the committee proceeded to make a general cleanup of the city's "bad men." A score or more of these "gentlemen" were rounded up and deported, with the warning never to return under penalty of death.

Prominent among the bad men were "Woolly" Karney, Billy Mulligan and "Yankee" Sullivan, bruisers of a most brutal type, and Charles P. Duane, at one time chief engineer of San Francisco's volunteer fire department.

Yankee Sullivan killed himself in his cell by severing an artery in his arm with a table knife. From the time of his arrest until he was found dying in his cell he was insane with fear, continually protesting that he was not guilty of murder, which he believed to be the charge against him. Taking into consideration his brutal character, his unholy life, his insane fear and protestations of innocence, there can hardly be a doubt that his hands had been steeped in the blood of his fellow man.

"Woolly" Karney ran a roadhouse on the Mission Road at the end of a bridge which spanned the great fresh water marsh that was there at that time. This roadhouse was the resort of the vilest gang of criminals and roughnecks on the American continent. It was suspected that more than one wayfarer had met his death there at the hands of "Woolly" and the gang. Tolled into the house, drugged, robbed and murdered, he would be thrown from the bridge into the mud and slush of the marsh, where the body quickly disappeared.

When the committee began raiding the gambling hells and other cesspools of evil, arresting and deporting the worst of their proprietors and habitués, scores of these undesirables having a wholesome regard for their liberty, and perhaps their necks, fled the city.

Thus was San Francisco made a clean city, where good people could safely live.

When the Vigilantes began their good work, Mr. Nugent in the *Herald* bitterly opposed them, denouncing them each morning in the most intemperate terms; stating that they were a mob of anarchists and rascals, banded together for the purpose of overthrowing the constituted authority of the city government, replacing it with mob law and violence. He also declared that hundreds of the members had joined it for the sole purpose of saving, as he termed it, "their own worthless hides."

The stand the *Herald* took in this, the darkest period of San Francisco's history, a period when she was groaning under a burden of lawlessness and misrule, caused a great wave of indignation to sweep over the city. A committee composed of some of the most influential citizens and business men waited upon Mr. Nugent, expostulating with him regarding his course, requesting him to cease his opposition, telling him his writings were not in the interests of good government, law and order. but in favor of anarchy and crime.

The next morning the *Herald* came out with the most violent editorial yet published, scoring the Vigilance Committee in the most abusive language, and applying to the gentlemen who had waited upon Nugent every opprobrious epithet he could think of. This article ended the *Herald* as a great newspaper. Every business man withdrew his advertisement, the list of subscribers dwindled to nothing, and within a few days the *Herald* shrank from a prosperous and influential journal to an eight-sheet pamphlet. This little sheet, filled with abuse of the Vigilantes, Nugent continued to publish for a short time, but, with influence, money and credit gone, deserted by friends, and shunned by almost every one, he at last struck his colors and quit. Of Mr. Nugent's subsequent life I know nothing, but I do know that his life as a newspaper man ended with that of the *Herald*. Steve, although disapproving Mr. Nugent's course, stayed with him till the end.

Steve Goes to Oregon

WHEN the *Herald* gave up the ghost in a losing fight against the Vigilance Committee, Steve Gillis was offered the position of editor and publisher of the *Occidental Messenger,* a newspaper published at Corvalis, Oregon.

After engaging L. P.— "Long Primer"— Hall, as associate editor, and Anthony Noltner, who afterward became mayor of Portland, as office boy and cook, Steve, in the early part of January, 1875, took passage on the steamship "Pacific," and left San Francisco to take possession of the "bag of gold" which he felt sure was waiting for him at the end of the rainbow at Corvalis.

Upon his arrival he and Hall were met by a committee of prominent Democrats whose chairman, a Mr. Avery, delivered an address of welcome, after which they were escorted to a hotel, where a fine chicken supper had been prepared. After partaking of the good things set before them, the evening was spent in speechmaking and expressions of mutual good will.

The next day Steve was taken over the town and made acquainted with the people, all of whom received him in a most friendly spirit, wishing him success and prosperity in his new undertaking. Steve now took hold of his work with great enthusiasm and energy and soon brought the *Messenger* to the front as one of the most ably conducted newspapers of the state. Its editorials were forcible and logical, and received favorable comment from not only the newspapers of Oregon, but the Democratic papers of California. Some of the leading articles were written by "Long Primer" Hall who, when sober, was one of the most brilliant writers on the coast, but unfortunately, he was addicted to periodical sprees and when on one of these he would perform some of the most foolish and ridiculous stunts imaginable.

One of these stunts he got off at Virginia City — the stronghold of the Whigs — where a great Whig mass meeting was being held. Hall was sent to report the proceedings of the meetings for the *Messenger,* which at the beginning of the campaign had been dubbed "Avery's Ox" by the Whig papers. Mr. Avery preceded Steve as publisher.

When the meeting was called to order it was

noticed that Hall was absent from his place among the other reporters. The principal speaker had been introduced and was in the midst of his highest oratorical flights with the crowd tensely listening, when Hall, with a big cowbell strapped around his neck, came charging among them, yelling at the top of his voice, "Here comes Avery's ox! Get out of my way, you ornery lot of Whigs, or I'll trample every last one of you into the ground."

Shouldering his way to the front he mounted the speaker's stand, where he continued his wild antics, stamping up and down the platform, shaking his great shaggy head, ringing his bell and bellowing his threats against the Whigs until, amidst a storm of yells and hisses, he was gathered in by a couple of huskies, and rushed howling and kicking to the cellar, where he was locked in until the next morning.

This was a number not down on the program, but a most effective one. It not only ended the meeting, but placed a club in the hands of the Whigs with which they belabored the Democrats throughout the state during the campaign.

This "wild bull" run of Hall's caused Steve great anger and humiliation and came near ending all relations between him and "Long Primer," but upon the latter's solemn promise to quit drink altogether, his escapade at Virginia City was overlooked and things went on as before.

Hall was a great lover of coffee, but he wanted it strong. He made poor Tony's life miserable by continually complaining of the dishwater he was giving him to drink. "Now, Steve," said he, "I am perfectly willing to let whisky alone if I have good, strong coffee to drink. When I have that I don't want whisky, but the kind of stuff Tony is making don't suit me at all. I wish you would tell him to put more coffee and less water in it."

Steve promised to do so. Now, at that time, Steve used neither tea nor coffee, so he concluded to play a joke on Hall. Going to a store, he bought a plug of Black Strap tobacco, and going into the kitchen the next morning, where Tony was preparing breakfast, he put the whole plug into the coffee, which was boiling on the stove.

While Tony was pouring the coffee, Hall was narrowly watching him, and something in the appearance of the coffee seemed to please him. Raising the cup to his lips, he took a swallow, then with a satisfied smile, emptied the cup at a single draught. Handing his cup back to Tony, he delightfully exclaimed, "By gum, Tony, my boy, you've struck it this time, sure. Now that's what I call good coffee;

it's the real thing, you bet. Fill 'er up again, Tony, fill 'er up." After drinking four cups of this delectable concoction, Hall rose from the table with a happy smile overspreading his face and went to his sanctum and wrote one of the finest articles on the topics of the day that ever came from his pen.

But, oh, how different it was with poor Tony! After drinking less than half a cup of the vile mixture, he arose from the table and staggered out of the room, a very sick boy, and lay down under a fig tree in the yard, where he stayed all that day, so sick and miserable that he wanted to die. For the next two days Tony was in such a weakened condition that he never left his bed and Steve had to take his place as cook and "devil." At breakfast the next morning after Hall had drunk the good coffee, he turned to Steve and said, "This is fairly good coffee, Steve, but, great Scott! it ain't a patching to that of Tony's yesterday morning. That was the most delicious coffee that I ever drank."

Steve continued publication of the *Messenger* until the end of the year, when, finding that it had not been a financial success, and having no hope of improvement ahead, quit and went back to his job on the *Alta* in San Francisco. Steve had made very many friends during his stay in Oregon. He was such a manly and friendly young fellow, so willing to help and do an act of kindness, so ready to cheer and comfort all those needing cheer and comfort, that he won the hearts of all with whom he came in contact. His course had been so clean and dignified as a newspaper man that even the writers who had opposed the principles he advocated expressed their regret that he was leaving Oregon, one Whig paper in Portland offering him a place on its editorial staff.

He stayed with his case on the *Alta* until the latter part of 1858, when he took over the business management of a newspaper at Tucson, Arizona, published by Judge Ned McGowan of vigilance committee fame, at a salary of $300 per month.

When Steve got this offer from McGowan another rainbow with the bag of gold at the end appeared and, of course, he went for it and went right away. When he arrived at Tucson he got the greatest shock he ever had. Previously his life had been passed among civilized, peaceable and educated people. Here in Tucson, it seemed to him, he had reached another world, to which all the bad men who lived on this planet had been transferred, each armed with rifle, pistol or bowie knife; each man watching his neighbor, ever ready to shoot or stab and kill, for any offense offered. A look, a word spoken in jest was sufficient. Steve was filled with horror and disgust, and finding that he was, as he told me, "up against it," threw up his job and returned to San Francisco.

When silver was discovered in Nevada in 1859, lured by stories of the fabulous wealth hidden in the Comstock and other great gold-silver bearing lodes, thousands and thousands of people went in on the great human tide that flowed into Nevada. Men from every nation and clime flocked to the feast. Every trade, profession and calling was represented — miners, lawyers, doctors, merchants and preachers gathered there, while hundreds of human vultures, scenting the prey from afar, flocked to the "Land of Promise." The city of Virginia, especially, was overrun by these tough citizens. They were known as "toughs" and the title did not displease them.

When Chief Meets Chief

AMONG them were about a dozen gunmen, each of whom had distinguished himself by killing one or more men. Chief of this crowd was Farner Peel, with six notches on the butt of his gun. This man, Peel, was of magnificent physique, six feet high and finely proportioned, a graduate of Harvard, highly cultured. He was a man of noble presence, courteous and gracious, and had he been clean of mind and soul would have been a welcome and honored member of any community and a desirable and popular member of the best society anywhere. Peel had never been the aggressor in any of his gun fights, but had always acted on the defensive, but every gunman who had started trouble with him had paid for it with his life.

Farner's reputation as a gun fighter had spread over the whole country, causing black envy to spring up in the minds of some of his rivals in other wild towns. The particular one of these with which this narrative has to do was "Austin Johnny," who had built up a great reputation in this line of sport in eastern Nevada. Johnny became obsessed with the idea that one chief was sufficient on this world of ours, and so he made up his mind to either kill Peel or be killed by him. With this laudable end in view he went to Virginia City and hunted up George Birdsall, told him his purpose and requested him to introduce him to Farner Peel. George did everything possible to induce him to forego his purpose, giving him a history of Peel's prowess as a gun fighter, telling him of his accuracy as a marksman.

"Why, Johnny," said George, "Farner can put all six balls from his gun into a circle an inch in diameter without lowering his arm, and flip a dime into the air and hit it before it reaches the ground."

"George, this talk's no good," Johnny replied, "I tell you I'm goin' to take the scalp of this Bully Boy, or he's goin' to take mine."

"Well, Johnny, your scalp will be the one taken, if you run up against Peel."

"Well," said Johnny, "if I'm goin' to die with my boots on, I guess I better have them polished," and he did have them polished then and there.

The two men walked on to Ben Irwin's saloon, where they found Farner laughing and talking with some of his friends. Approaching him, George introduced the two men. After shaking hands, Johnny asked Peel and his friends to drink. Then, he, addressing Farner, said, "Mr. Peel, you are chief over this here neck of the woods, and I am chief of the rest of the world. There is not room on this world for two chiefs, so one of us has got to pass in his chips."

"Why, my friend," said Peel, "I have no desire to be chief anywhere, or of anything. I am one who never seeks trouble. It is true that I have been compelled to defend my life against the attacks of desperate men but I have never been the aggressor, and it has been to me a source of sorrow that I was ever forced to take the life of a fellow man. Any of the gentlemen in this room will vouch for the truth of what I have told you. You have a wrong idea of my character. You have no cause of quarrel with me. Come, have a drink, shake hands and be friends."

"This talk of yours, Peel, don't cut any ice with me. I tell you," said Johnny, throwing his hand to his pistol pocket, "that one of us two has got to die!"

"Well," said Farner, "if nothing will satisfy you but gunplay with me, don't let us pull it off here. This room is crowded with people who have no part or parcel in this affair, and maybe one or more might be hurt or killed. Let us go out to the street where the life of no one else will be endangered."

So it was agreed upon, Johnny going backwards through the saloon toward the door with Peel close behind him. When Johnny reached the opposite side of the street he squatted and was taking aim with his pistol on his left arm, when Peel, standing in the doorway of the saloon, raised his pistol and fired, shooting Johnny squarely between the eyes and killing him instantly. Thus ended the activities of the "Big Chief" from Austin.

Two or three of those who had witnessed the shooting went where he lay, and, finding he was dead, lifted the body and carried it into the saloon. When Farner Peel reëntered the saloon he said to Birdsall, "George, what, in the name of all that is evil, made you bring that boy here to have trouble with me? You surely knew that he had no chance to win out in a game like that with me, and you should have kept him away."

"Farner," said George, "I did all I could and used every argument I could think of to prevent his coming, but it did no good. I even threatened to lock him up. When I made that threat he flew into a rage and I thought he was going to turn his gun on me, so I gave up and brought him along."

"Oh, well," said Peel, "it can't be helped now, but I am very sorry that it happened. That poor fellow knew nothing about handling a gun, and it was like shooting a baby, but what could I do? He had his gun trained on me, had it gone off, my body, instead of his, might be lying there, and I almost wish it was. George, you go up to Brown's and tell him to come and prepare the body for burial. Tell him to spare no expense, but to give him the best he has in the shop, and I will pay for everything."

Brown dressed the body in a fine burial robe and placed it in a magnificent casket, elegantly upholstered, and "Austin Johnny" was ready for the grave. Johnny's body lay in state from Friday until Sunday in Irwin's saloon. During those two days seven bartenders were kept busy handing refreshments to the thirsty hundreds who visited the saloon to look for the last time upon the dead face. The remains of this gun fighter were escorted to the cemetery by the largest funeral procession ever known in Nevada. The doors of every saloon in Virginia City, draped in mourning, were closed, and an air of sadness seemed to pervade the whole community. A brass band, playing mournful music, led the procession to the cemetery where the body of "Austin Johnny" was laid to rest.

Pete Larkin, another gunman who was afterwards hanged for murder, was master of ceremonies at the grave, and I wish that I could give his remarks as he made them, but that is a task that only a Mark Twain could perform. A stranger viewing this great demonstration would have thought that some great statesman or benefactor of his race was being honored and mourned by a grief-stricken community. After Johnny's body was laid in the grave, the people marched back to the city to the tune of "When Johnny Comes Marching Home Again," played by the band, and the show was over.

Enter Samuel Clemens

In 1861 Joe Goodman and Dagget McCarthy established the *Territorial Enterprise* at Virginia City. Steve, my brother, was foreman of the composing room.

Mark Twain had been writing news items for the *Enterprise* for something like a year when Goodman offered him a place as reporter on the paper. One night, shortly after this offer had been made, while Joe and Dagget were sitting in the editorial room enjoying a quiet smoke waiting for "copy" from the composing room, a young man with a great bushy shock of brown hair entered, and a long-drawling voice said, "Does either of you gentlemen happen to be Mr. J. T. Goodman?"

"I am Joe Goodman," said Joe. "Can I do anything for you?"

"Mr. Goodman, if you will look me over I think you will observe that the clothes I have on would not be suitable ones to wear at a fashionable tea party, also that a haircut and a shave, together with a new hat, would greatly improve my personal appearance. A bath, too, would feel mighty good to me, and a steak or a dish of ham and eggs would be very satisfying."

"My friend," said Joe, with a broad smile, "if you will just look this room over, I think you will observe that it is neither a clothing store, nor a barber shop. You will find, also, that there is no cuisine connected with it. I am sorry for you, but you have come to the wrong place, and will have to apply elsewhere."

The stranger, presenting a letter, said, "Mr. Goodman, if you will read this letter, I think you will agree with me that my reception has not been as cordial as I was led to expect it would be."

"Well, by the great Hornspoon," exclaimed Joe, dropping the letter, "Dagget, this is Sam Clemens, and I'll be darned if he hasn't put one over on us this time, all right. Why didn't you say who you were in the first place, Sam, instead of beating around the bush in the way you did? But never mind, we're mighty glad you are here. Sit down. Dagget, step to the window and tell Steve to come here."

Dagget went to the window connecting the editorial and composing rooms and at the top of his voice shouted, "Steve Gillis, you are wanted in the editorial room." Steve lost no time in answering this summons, but came on the run. On entering, he saw Joe sitting at the table with Dagget standing just behind him, while on the opposite side stood an unshaven, roughly clad individual shaking his bushy head, and, as Steve thought, frowning at Joe. Appearances seemed to call for action, and Steve began to roll up his sleeves.

"There is no trouble, Steve. I just called you in to make you acquainted with Sam Clemens. Mr. Clemens, this is our foreman, Steve Gillis."

With one of his hearty laughs, Steve started, with extended hand, across the room, saying, "Why, Mr. Clemens, I am surely glad to see you, although you don't look a bit like the Sam Clemens pictured in my mind."

"Mr. Gillis," said Sam, "I don't often cotton to a man on first acquaintance, but I do cotton to you, right here and now, and I know we're going to be friends right from the start."

And so it proved to be. There was never a truer or closer friendship between two people than that between Sam and Steve.

"Steve," said Joe, "Sam has walked all the way from Aurora here and he thinks he needs some repairs. You go down to the office with him and tell Charlie Bicknell to let him have the money he needs, and then get Mike's boy to pilot him to the places he wishes to patronize."

"Mike's boy, nothing," said Steve. "Just wait till I put Joe Harlow in charge of the forms, and I'll go myself."

After Sam had provided himself with a suit of clothes, a haircut, shave and bath, he and Steve went to the "White House," where they had a hearty supper. Then they returned to the office, where Sam kept them until morning telling, in his droll way, of his experiences from Aurora to Virginia City.

The morning after his arrival Clemens took charge of the local columns of the *Enterprise*. Associated with him was Dan McQuill, another writer widely known all over the coast as a humorist of exceptional ability. These two men kept the people along the Comstock continually laughing.

In December, 1862, Steve went from Virginia City to San Francisco and took a case on the *Morning Call*. Here he continued to work until September, 1864. After Steve left, Sam became dispirited and did not seem at all like himself. He had lost his friend and could not be comforted. He stood his loneliness for two weeks and then resigned his position and followed Steve to San Francisco, saying that he knew Steve would be sure to get into some kind of trouble, and that it was his duty to get him out of it. This prophecy of Sam's was fulfilled.

The Fight That Made Mark Twain Famous

ONE night when on his way home from work, in passing the saloon of "Big Jim" Casey, on Howard Street, Steve saw a big ruffian unmercifully beating a little chap who had no more show with Casey than a rat would have with a terrier. Now Steve was ever ready to help the weak against the strong, and the spectacle he was witnessing irresistibly called upon him to hasten to the relief of the little fellow in distress. So, without hesitation, he entered the saloon and took part in the fray. When Steve interfered in the fight, Casey let the little man go, locked the door and put the key in his pocket. Then turning to Steve, he said, "Now, Mister, as you have butted in without being asked, I'll finish the job on you." And with these words he went at Steve with a rush.

When Casey locked the door Steve stepped over to the bar and there stood waiting the onslaught; and when within striking distance, smashed him over the head with a big beer pitcher standing on the bar. This blow was a "down went McGinty" one for Casey, and he fell heavily to the floor, all the fight knocked out of him.

Now was the time for Steve to make his getaway, but the key being in Casey's pocket he had no way of leaving the saloon, so he stood quietly and waited until Billy Blitz and another officer appeared upon the scene. Blitz at once stooped to examine Casey. Finding him unconscious, with blood streaming from his scalp, he summoned an ambulance and had him conveyed to the county hospital. He then arrested Steve and took him to the police station, where a charge of assault and battery was preferred against him.

Mark Twain, being summoned, went on Steve's bond in the sum of five hundred dollars, and he was released. When the two friends left the Hall of Justice they walked along in silence for a short distance, with Sam in the lead, shaking his head and muttering to himself. Sam's aloofness on this occasion was so unusual that Steve couldn't comprehend it, so at last he hailed him with, "Hold on, Sam, don't be in such a hurry. What's the matter with you, anyhow?"

"I'm mad, that's what's the matter, if you want to know, you confounded jackass. I am disgusted with you and don't care to have any talk with you at all."

"Why, Sam, I don't see that I have done anything so awful as to cause you to feel like that towards me."

"Oh, you don't, don't you? You didn't deliberately get into a mixup with Big Jim Casey over an affair that didn't concern you, did you? Why didn't you go on home last night, like a decent man? If you had, none of this trouble would have occurred. You would not now be booked at the city prison as a law breaker and a criminal and my name would not be registered there, as the friend and bondsman of a scrapping idiot. Instead of acting like a sensible person and attending to your own affairs, you had to go and jump into that dirty row. Oh! you have thrown the fat in the fire this' time, young man."

"I say, Sam, you are going too far. You must let up on that kind of talk! I did no more than any man with the least humanity in his heart would have done in the same circumstances. When I got as far as Casey's saloon, on my way home, I saw the big man literally wiping up the floor with a little shrimp of a man, who was no more able to contend with him than a sparrow would be with a cat. He was just beating that poor little chap to a frazzle. Sam, I couldn't stand for anything like that, and I had to stop Casey. I had no intention of getting into a fight with him. I only meant to make him let up on the little man and then go home. When I interfered he quit beating the little fellow, but, instead of quieting down and behaving himself like a decent man, he locked the door and put the key in his pocket, came for me with his head down, like a mad bull. I saw he intended to butt me in the stomach, so I grabbed the pitcher from the bar, and when he got within reach, smashed him over the head with it. No, Sam, that is all there is to this assault and battery business. He is the one who made the assault and I, in self-defense, stopped him with the pitcher."

"Yes, you stopped him all right, and I guess all that you have told me is true, but when this case comes to trial, if it ever does, what you have just told me is not going to help you, for the reason that you have no witness to substantiate your story."

"Well, Sam, here we are at the house. Let's stop talking about it till later on and, for goodness sake, don't let mother and the girls know anything about it."

"Haven't you got brains enough in that thick head of yours to know that a policeman couldn't come here at 2 o'clock in the morning and snake me off to the station house without their knowing that you were in trouble? You should have thought of your mother and sisters before you went into that rotten row. Steve Gillis, you make me sick. Come on and let's get to bed."

The next morning as they were on their way to the *Call* office they met Officer Blitz.

"Billy," said Sam, "have you heard anything of Casey's condition this morning?"

"Yes, I have, and it is a mighty serious one. His scalp is frightfully cut, and at 10 o'clock this morning he was raving in delirium. Steve, you hit Jim a hell of a lick with that pitcher, and I wouldn't wonder if he's knocked out for good. I am sorry, Steve, but I'm afraid you are going to have a heap of trouble getting out of this business."

After Blitz left them, Sam and Steve continued on their way. "Steve, from what Billy has just told us, it looks like Jim Casey has a good chance to die. If he does, you are going to have the devil's own time getting out of this rotten scrape. The only way I see that you can get out of it is to skip out of San Francisco and go back to your case on the *Enterprise*."

"And when you get out of here, what's going to happen to me, do you think?" asked Clemens. "When the case is called for trial and you fail to appear, your bonds will be declared forfeited. Dave Louderbach will then hale me into court and try to make me put up five hundred dollars for being fool enough to go on your bonds. I'll either have to pay the money or go to jail, and I don't want to do either."

"Say, Sam," said Steve, "if I have to go back to Virginia City, and I guess I had better, you go up to my two brothers on Jackass Hill and stay with them until this thing blows over. They will be delighted to have you with them. It will be a splendid vacation and outing for you, and you will have the time of your life. It won't interfere with your engagements with the papers for which you are writing here, and you will be able to pick up a lot of things that will help you as a writer."

Mark Goes to Jackass Hill

So it was arranged. Steve left the next day for Virginia City, where he resumed work as a compositor. A few days after Steve's departure Sam left for Jackass Hill, where he was the guest of myself and my brother Jim for something like five months.

During the time of his visit to us, Sam Clemens never left Jackass Hill, except for a short time spent at Angels' Camp, Calaveras County, when he gathered data for the "Jumping Frog of Calaveras" and "Roughing It." Steve became a prophet when he said to him, "You will be able to pick up lots of things that will help you as a writer." With the publication of these two books his career as the greatest humorist of his time really began.

Shortly after his return to San Francisco from his sojourn on the "Hill," Sam was sent with a party of United States surveyors to the Sandwich Islands as correspondent of the *Alta California*. While on this trip he wrote a series of letters for that paper which made the people laugh and cry all over the world. Returning to California, he lectured in all the principal cities of the coast on his experiences in the Islands and was enthusiastically received everywhere.

Shortly after completing his course of lectures on the Sandwich Islands he was sent by James Gordon Bennett of the *New York Herald* to the Holy Land as the special correspondent of the *Herald*. He accompanied a party of eminent divines, who went there on a voyage of discovery and research among the sacred places of Palestine. It was while on this journey that he collected the data for what many regard as his greatest work — "The Innocents Abroad." Coming back from the Holy Land he delivered a series of lectures on the country.

Mark Twain's fame was now assured and he was anchored in the hearts of the people of the whole world. Opportunity knocked at his door when Steve had his famous fight with Big Jim Casey. Had that fight not occurred, he would not have gone to Jackass Hill, the "Jumping Frog" and "Roughing It" would not have been written. No "Letters From the Islands" would have made the world laugh and the story of "The Innocents Abroad" would not have been told. We must conclude that Mark Twain owed his quick transition from a newspaper correspondent to the greatest humorist of his time to that mixup between Steve and Casey that day in September, 1864.

Mark Twain and the Burning Fuse

I WAS interviewed on one occasion by a gentleman from the East, who, after I told him all that I thought would interest him, asked, "Is that shaft still open where Mark came near losing his life by the explosion of a blast of dynamite because the man tending the hoist went to sleep?"

"That is a story," said I, "that I never heard. Tell me about it."

"Well, Mark drilled a hole in the bottom of the shaft, loaded it with a heavy charge of dynamite, lighted the fuse and rang for the skip. After waiting several minutes without receiving any response to his signal, he rang again. The result was the same. By this time the fuse had burned pretty close to the top of the hole. Mark then concluded that further delay would be dangerous, so he took his knife from his pocket and quietly whetting it on his boot, cut the fuse when the fire was within half an inch of the mouth of the hole. Ringing the bell again, he sat calmly down, lit his pipe and patiently waited for his partner to hoist him to the surface. In about half an hour he heard a sound like some one stepping on the platform, and, raising his voice, called, 'Oh, Dick!'

"An answer came back, 'Hello, down there. Anything wanted?'

"'Well, yes, if it is convenient I would like to be hoisted to the top.'

"When the skip reached the bottom Mark rang three bells — meaning 'man aboard, go slow' — and, relighting the fuse, stepped into the skip and rang the signal to hoist. When he arrived at the surface, his partner said to him, 'Why didn't you ring the bell, Sam, instead of calling?'

"'I did ring the bell, but it brought no results. Where have you been?'

"'Nowhere. I've been right here all the time. Why do you ask?'

"'Oh, I just wanted to know, that's all. But your failure to answer my signals to hoist came very near causing you to lose your partner, and me, my life. How did it happen that you failed to hear the bell?'

"'Well, Sam, you know that it was nearly morning when I got home from the dance. I was feeling pretty rocky, and after I lowered you down the shaft

I laid down with my coat for a pillow to take a little rest. I didn't intend to, but I must have gone to sleep."

" 'Now, Dick, I don't want you to think for a moment that I have any hard feelings towards you, but your going to sleep on this occasion came very near causing my death, but I am as much alive as ever, so we'll just forget it and go to dinner.' "

If such an experience as the above could have possibly have come to Sam Clemens, instead of talking the incident over with Dick in a calm and friendly way, he would probably have grabbed a drill, pick handle or any other weapon coming handy and brained him with it.

Clemens' One Mining Venture

THE nearest approach to any work that Mark Twain ever did at mining was when he became my partner at one time for about two weeks. One day when on my way home from Sonora I took a short cut across a parcel of land from which the surface dirt had been washed by the placer miners some years before. While walking over this ground I came to a spot where the croppings of a reef of very fine mineral slate had been uncovered, and upon a closer examination, discovered a small quartz vein with a clay casing running through this body of slate. The chances for finding a "pocket" here looked mighty good to me, and I determined to return the next day and give it a tryout. Putting a piece of quartz in my pocket, I continued on my way home. On entering the cabin I found Sam just sitting down to supper.

"Hello, Billy," said he, "you are just in time for the feast. I won't sing the praises of the bacon, but

I'll bet your stomach never entertained slapjacks like these in your life."

After supper I handed him the quartz and said, "Sam, the chances for finding a pocket where I got that rock look mighty good to me. Go over with me in the morning and we'll go snooks in anything we find."

"All right, if you are willing to take me in, knowing that I know nothing about pocket mining, I will go in with you and we'll dig out a million or two."

Next morning, on arriving on the ground, I said to him, "Sam, you sit down and rest while I shovel off some of this clay and see what the vein looks like where it is in place."

"All right, I'll take a smoke while you are doing that."

I took my pick and began digging. I had been at work only a short time when appearances were so favorable that I concluded to "take a pan." So, taking my crowbar, I began gouging the vein. I had filled the pan about half full when I saw color of gold.

"Sam," I called, "I guess we've struck it. There's gold in sight."

"That's bully," said he, and coming over to me,

sat down and watched me gouging, the while scratching the clay and quartz in the pan over with a small stick. When I had the pan sufficiently full, I shoved it over to him and said, "Here, Sam, take this over to the panhole and wash it, while I take out another."

"Billy," said he, "I wouldn't puddle in that confounded clay for all the gold in Tuolumne County. You pan it; I don't want any of it."

"Very well, Sam, just put it in the pan-hole to soak till I get out this other pan and I'll wash it myself." The result from that first pan was about five dollars in gold, and by quitting time I had panned out about one hundred dollars' worth. So it went for the next ten days, I doing the work and Sam superintending. At the expiration of that time I had extracted all the gold from the pocket, which amounted to about seven hundred dollars. When I received the returns from the mint I proffered Sam one half of the money as his share as my partner in our mining venture. He refused to take any of it, saying, "The knowledge of mining I acquired while we were taking out that pocket, and the pleasure it gave me, is a better equivalent for my time and labor than that little dab of money."

Fun on the Hill

GETTING into Tuttletown at a rather late hour one night on my way home from Sonora, I found a party of half a dozen young men who had been serenading their lady friends in the neighborhood. I suggested that they go with me to Jackass Hill and end the night's program with a serenade to Mark Twain. They readily fell in with my suggestion and we climbed the hill together, and, after our chief musician had tuned up his "old banjo," lined up under Mark's window, and opened up with "Oh. Darkies hab you seen Ole Massa?"

We had finished this song and "Happy Land O' Canaan" and were well under way with "I'se Gwine to de Shuckin," when that window went up with a bang, and an angry, rasping voice snarled out, "What do you lot of yapping coyotes mean by disturbing the peace and quiet of the respectable people on the hill with that infernal yowling you're doing out there? Get away from this window, you drunken loafers, and go off to that shuckin' you're howling about, and go right now."

This rude reception, it is needless to say, put an abrupt ending to our serenade and my companions left the hill on the double quick. On entering the cabin I found Mark sitting on the side of the bed, cramming his pipe with "Bull Durham" tobacco. "Hello, Sam," said I, "going to have a smoke?"

At my salutation he looked at me with an ugly scowl and greeted me with, "Billy, how did you come to get drunk to-night, and bring that gang of low-down rowdies on the hill, to make the night hideous with their horrible racket? Up to this time I have regarded you as a well-behaved, decent young fellow with instincts somewhat approaching those of a gentleman but I have been wakened from that dream to-night to find you nothing but a common, wine guzzling hoodlum."

"Why, Sam," said I, "I'm very sorry that you regard me in such a light, and if you will think a moment, I believe that you will admit that you are doing me an injustice. I had no thought, nor did the other boys, of giving offense by singing for you to-night. Our only intention or desire was to give you pleasure, for you know, Sam, that upon the occasion of our visit to the young ladies of French Flat you told me that 'music had charms to soothe the savage' and advised me to cultivate — if I had any — my talent for it, and I was led to believe from what you said that you loved music."

"Music! Great Scott, do you call that music? Why, the bleatings of Carrington's band of goats are like strains of melody from the heavenly choir compared to that horrible catawailing you were doing there. Music, indeed, huh."

"Oh, well, Sam," said I, "I see now that I had the wrong coon up a tree, and I ask your pardon. Come, shake hands, and be my friend again, and I promise that there will be no repetition of to-night's work."

"I'll have to do some serious thinking before I give you my friendship again, and as to shaking hands with you while in your present maudlin condition, I tell you, Billy Gillis, I won't do it. Take off your clothes and go to bed, and try to sleep off your drunken debauch. Get over, as far as you can, to the other side of the bed, and turn your face to the wall, for the fumes of that infernal stuff you call wine that you have loaded up with, make me sick at my stomach."

"All right, Sam. Good night."

Shortly after Sam returned to the "Hill" from

Angels' Camp, the boys instituted a "Hospital for the Insane," on Jackass Hill. Our officers consisted of a board of directors and a resident physician, the rest of us constituting the attendants and patients. The "doctor" held office for one week, then gave place to one of the others. When Sam's turn came he sent in something like the following report:

"To the honorable board of directors of the Hospital for the Insane on Jackass Hill.

"Gentlemen: I have the honor to present to your honorable body the following report of my administration of the affairs of this institution for the week ending February —, 1865. I am happy to state that, with one exception, the inmates under my care are rapidly approaching complete recovery and I am greatly encouraged to believe that they will soon be in full possession of their mental faculties. The exception, noted above, is a young man named James N. Gillis. In all respects save one this patient's mind is in perfectly normal state. He is a very companionly young fellow, and tells some fairly humorous stories, and it is sad to know that this young man, who would otherwise be a useful member of society, is hopelessly insane, but such, I am sorry to say, is the truth. He is laboring under the hallucination that he is the greatest pocket miner on earth; that he can save more gold in panning out and can better appraise the value of a 'clean up' before weighing than any other man; and that he is the only miner having a perfect knowledge of gold-bearing ledges and formations. He is a fairly good pocket miner and knows a gold nugget from a brass door knob, but there are a dozen boys on the hill who can give him cards and spades and beat him at the game."

This report of Sam's was received by the boys with shouts of merriment, Jim joining in the laugh at his own expense. The next week Jim was the doctor. His report to the "Board of Directors" read like this:

"One of the most pitiful cases of insanity that has ever fallen under my observation is that of a young man named Samuel L. Clemens, who was committed to this hospital on the thirteenth day of last month, from Angels' Camp, Calaveras County. In my conversation with him on the day after he came under my care he seemed to be perfectly rational and in full possession of his faculties. I was greatly encouraged to hope that after a short treatment he would be restored to his friends and society and again take his place as a useful worker in the affairs of the world. It was but a short time, however, when this hope was rudely shattered, and I found that he was hopelessly insane. He has, for the past three years, been associated with newspaper men of rare literary ability. He is obsessed with the idea that they are the spokes of a wheel and he the hub around which they revolve. He has a mania for story writing, and is at the present time engaged in writing 'The Jumping Frog of Calaveras,' which he imagines will cause his name to be handed down to posterity from generation to generation as the greatest humorist of all time. This great story of his is nothing but a lot of silly drivel about a warty old toad that he was told by some joker in Angels' Camp. Every evening when the inmates are together in the living room, he takes up his manuscript and reads to them a page or two of the story and then winds up with, 'It's no wonder, that darned frog couldn't jump worth shucks, boys, filled up with shot like he was.' Then he will chuckle to himself and murmur about 'copyrights' and 'royalties.' If this was the only trouble with Mark Twain, as he dubs himself in his stories, there would be a reasonable hope of the ultimate restoration of his mentality, but the one great hallucination that will forever bar him from the 'busy walks of life' is that he was at one time a pilot on one of the great Mississippi River packets which plied between St. Louis and New Orleans. He delights to tell his experiences while navigating one of those big boats on the great river, how his knowledge of its currents and bends, its shoals and eddies, and the dangerous snags along its banks together with his quick manipulation of the wheel saved his boat and the lives of his passengers. Poor Mark! His nearest approach to being a pilot on the river was when he handled the big steering paddle of a flatboat, freighted with apples from Ohio, which were peddled in towns along the river."

This report of "Doctor" Jim was received by all the boys, except Sam, with roars of laughter and applause, but nothing of the kind came from him. On the contrary he sat through its reading livid with rage. After the merriment had somewhat subsided he opened on the boys with "You lot of laughing jackals! You think the rotten hogwash read to you by that empty-headed idiot, Jim Gillis, is mighty funny, don't you? I have had the impression up to to-night that you fellows had a few brains stowed away somewhere in your coconut heads, but now I see my mistake, and find that I have been associating with a crowd of ignorant, grinning apes, instead of intelligent human beings. I appreciate a joke, and love fun as much as any boy in the world, but when a lot of rotten stuff like Jim Gillis' funny hash is pulled off on me I am ready to cry quits. I do quit right now, and will have nothing more to do with your fool funny business."

Sam did like fun, but not when the fun was at his expense. There was never another meeting of the the directors and inmates of the "hospital," Jim's report causing the institution to close for good. For a few days after this the boys kept Sam at fever heat with anger by asking him such questions as, "Say, Sam, how many barrels of apples could you load on that flat-bottomed scow?" "Sam, how long did it take you to float that flatboat from St. Louis down to New Orleans?"

This chaffing so exasperated Sam that Jim requested the boys to stop it, fearing that he would leave the Hill altogether. It was but a short time after this that he did leave and return to San Francisco, greatly to the regret of all the jolly good fellows whose esteem and friendship he had won while with them on the hill.

Just previous to his departure for Honolulu I received a letter from Sam, which ended, as nearly as I can now remember, as follows:

"I am leaving San Francisco in a short time for the Sandwich Islands with a party of U. S. surveyors, as special correspondent for the *Alta California*.

As in the course of human events we may not meet again, I will unburden my conscience of a load it has been carrying ever since the night of the serenade you and your band of troubadours attempted to give me. When you came into the cabin after I had scared the other boys off the Hill, I was in a mighty ugly mood and I wanted just the chance you gave me to vent my spleen on somebody or something. I called you some pretty hard names, which I knew at the time were undeserved, and accused you of high crimes and misdemeanors of which I knew you were not guilty. I wanted to ask your pardon the next morning at breakfast, but courage failed me and I put off doing so to a more 'convenient season.' That season has now arrived, and I do ask you to forgive me. Tell the boys that I am often with them in my dreams, and that when I return to the city I will come back to them once more on Old Jackass, if I can possibly arrange to do so."

Sam never visited the Hill again. After his return to California his time was so fully occupied in preparation for his lecturing tour on his trip to the Islands that he found it impossible. It will be remarked that in this narrative I have seldom used his nom de guerre of Mark Twain. He has always been to me, and to all his other friends and companions who roughed it with him in the days of yore, not the great humorist whose name is a household word over the whole world, but just plain Sam — good old Sam — one of the boys.

Conditions Along the Comstock

THE Great Comstock Lode passes the upper side of Virginia City in a north and south direction along the eastern slope of Mount Davidson, and through Gold Hill to the southward, and the portion of that famous lode included within the limits of these two cities, contained, without doubt, the richest known silver mines in the world. As early as the fall of 1859 a few miners, principally emigrants from across the Plains, were working surface gold diggings in Gold Canyon, and making fair wages, say from half an ounce to an ounce per day, with pans and rockers. The surpassingly rich gold fields of California were, however, the all-absorbing idea; therefore, few cared to stop and mine in this uninviting locality, but the general desire of all emigrants was to push forward to the golden Land of Promise on the other side. Earlier some little prospecting was done in neighboring ravines and gulches; yet it was not until 1857 that gold mining was prosecuted to any extent in

Six Mile Canyon, the first large ravine to the northwest of Virginia City. Gradually the miners prospected and worked their way along up Gold Canyon and above Devil's Gate, until in the winter of 1858 and spring of 1859, rich surface diggings were developed on the westerly side of the ravine. It was merely the decomposed croppings of the Comstock Lode. But the miners had but little idea of quartz mining at that time, much less of silver ore. Indeed, although the precious dust washed out by them contained a large proportion of silver, yet the idea of being actually at work on the surface of the richest silver mine in the world did not seem to enter their heads, and it was not until some months after discovery of silver ore had been made by the miners of Six Mile Canyon, near Cedar Hill, that a silver light flashed over the country, causing a wild rush hither, especially from California. The location of the first silver discovery and other advantages caused Virginia City to become the principal gathering point of the prospecting, speculative multitude; and she grew and flourished in exceeding prosperity.

Where, but yesterday, as it were, there had been nothing but a wilderness of sage brush and sand, a great city sprung up; populated by people of every nationality, where every branch of industry was represented, from the "Man of God with a message of good will to men," to the professional gambler, saloon keeper and dead beat.

The surrounding hills were dotted with shacks and tents, with here and there a more pretentious building. These latter buildings, for the most part, were occupied as saloons, gambling hells and dance halls.

At this time Virginia City had no organized government, consequently every man became "a law unto himself." Might was right, the strongest man was the best man and he took what he wanted. Robberies and acts of violence were of daily occurrence, and the crack of a pistol, at night, told the people that there would be "a man for breakfast" the next morning. The holdup man made no secret of his work, and openly disposed of his booty.

These conditions finally became so bad that the law-loving people of the city got together and organized a "Vigilance Committee" to protect the community.

A short time subsequent to the organization of the Committee, a young devil named John Hefferman, waylaid and wantonly murdered an old man for what little money he had on his person, then boldly proceeded to a gambling den and "blew it in" at faro. He was arrested, tried, convicted and hanged by the Vigilantes within twelve hours after the perpetration of his dastardly deed. The summary execution of Hefferman had a wholesome effect upon the conditions in Virginia City in so far as open acts of violence and crime were concerned, but did not, by any means, end the activities of the criminal element of the city. Murders and robberies still occurred with great frequency, but these human hyenas became more careful in covering their tracks and removing the evidences of their crimes.

The "Divide" between Virginia City and Gold Hill was an ideal locality for the deadly purposes of the holdup man. Here he could rob, and, meeting resistance, murder his victim, throw the body into an abandoned prospect hole and hide all evidence of his guilt. There were a score of miners, who after receiving their wages on pay day, never reached their habitations but disappeared as completely as though they had never been born. There is no doubt that they met their death at the hands of the inhuman wretches who waylaid them on the "Divide." (The Divide is the place where Mark Twain was held up by his friends, an account of which I will give further along.)

When the Government of Virginia City was

organized, an election was called, at which were selected a mayor, city attorney, chief of police, chief engineer of the fire department, one justice of the peace and a board of aldermen. As soon as these gentlemen were inducted the Board of Aldermen formulated a set of ordinances under which the affairs of the city were to be administered. These ordinances constituted an excellent code, could they have been enforced; but under the existing conditions, they had as well been written on sand, so far as the major criminals were concerned.

Petty thieves, vagrants and drunks were arrested and duly punished; the jails were crowded with this class, but owing to the fact that one rough element had so perfect an organization, it was not possible to convict members of that "Brotherhood."

It would be an injustice to these men to class them as criminals in the common acceptance of the word. Most of them were good citizens in every way except in the matter of gun play. They were a free-hearted, fun-loving lot of fellows, ever ready to help where help was needed, and to generously contribute to everything that stood for the public good, but they were all "gun-men," quick to resent an insult or to avenge a wrong with a bullet.

They were also guilty of many outrages in the

way of rough practical jokes, which often resulted in permanent injury to, and in two cases, the death of, their victims. Consequently this "Fraternity" of jolly good fellows was a constant menace and terror to the law-abiding and peace-loving citizens of Virginia City. The law was powerless to punish these men, for whenever one of them killed, no matter where or under what circumstances, and was brought to trial, he always had a number of witnesses to prove that he had killed in defense of his own life. The jury would, of course, render a verdict of acquittal, and he would walk out of the court room a free and, under the law, an innocent man.

Lee Invites His Old Schoolmate to Visit Him at Virginia City

IN the summer of 1868, one of this group of toughs, a fellow named Lee, learning that one of his boyhood friends and schoolmates was living at Helena, Montana, wrote him a letter inviting him to come to Virginia City on a certain day. When the stage from Reno arrived, Jim found his old schoolmate, accompanied by a score of companions, waiting to greet him. His reception by these people was so cordial, and their expressions of good will rang so true, that Jim's heart warmed to them at once, and he felt that his lines had been cast in pleasant places.

During the next ten days the town was painted red by these two men and their friends, making the rounds of all the saloons, gambling hells and dens of vice, of Virginia City and Gold Hill. When this palled upon them, filled and crazed with booze, they would go upon the streets at night and indulge in any sport that would afford them amusement; firing

He then asked to be shown to a place of entertainment, where he could get food and a place to sleep. When he made this request, Riff Williams, one of the "boys," stepped to his side and taking his arm, said, "Lean upon me, Lord, and I will lead Thee to the House of Partridge, the Publican, where Thou wilt find shelter and rest."

Followed by a noisy, laughing crowd, they proceeded up C Street to the International Hotel. Upon entering the office, Riff led his companion to the desk, and addressing the proprietor said:

"Here, Host, is a traveler, who fain would tarry with thee for a season; see that thou givest him freely of thy substance, that it may be well with thee."

"Very well," said the proprietor, who seemed to instantly grasp the situation, "I will do everything possible for his cheer and comfort." Then turning to the stranger, said, "How can I serve you, Sir?"

"I am weary and hungry, and would have food and rest," was the answer.

"Then come with me and you shall have both." Mr. Partridge then conducted him to a comfortable room and had him amply provided with food. After having performed these good offiicers for his guest, he re-entered the office, and extending his hand to Riff, said,

"I most heartily thank you, Mr. Williams, for piloting that young man to this hotel to-night, and I assure you I shall not forget your kindness."

"I say, Mr. Partridge," said Riff, "what's the idea? Do you mean to tell me that you're goin' to keep that nut, here in the hotel, all night."

"I certainly do, Riff, and not only to-night, but every night while he is in the city."

"Well, I'm damn, if that don't beat me. Why, me and the boys was going to entertain him ourselves with a little run around town between now and morning — showing him the sights and all that, and now, you are spoiling the fun."

"Well, you fellows are not going to have any horse play with that poor boy to-night. He is here now, and here he is going to stay."

"Say, boys, I guess that cans the fun," said Riff, "so I reckon we better beat it." And out they went.

The next morning the young fellow started on his canvass of the city, informing those with whom he came in contact, as to his identity, his object in coming to Virginia City, and receiving their contributions. He had visited both of the newspaper offices, where he had met with a kindly reception, and was coming out from the hardware store of Gilling & Mott, when Lee, who had been looking for their guns into the air, singing ribald songs, and playing rough practical jokes on any one who was so unfortunate as to meet them.

Their debauch ended in the killing of Jim, the Montana man, by his friend Lee. Returning from a raid of the Chinese quarter on the last night of their celebration, they entered the saloon of Doyle and Goodman, and lined up at the bar to take a fresh supply of booze, every one in the room being invited to drink. After telling their appreciative audience of their visit to Chinatown, of the fun they had had, and the jokes they had played on the Celestials, Lee and his friend became reminiscent, and began recounting incidents of their school-boy days.

One of these stories regarding a fight over a game of marbles caused a bitter quarrel between them, which waxed hotter and hotter, until their friends had to interfere to prevent them coming to blows and, perhaps, gun-play. After peace was re-established, the two men, apparently reconciled, left the saloon arm in arm. After reaching the street, however, the quarrel was renewed and resulted in a rough and tumble fight, in which the Montana man gave Lee such a drubbing that he had to be carried to his home by two of his friends.

The next night, smarting under a sense of injury

and defeat, still crazed with booze, Lee armed himself and started on a hunt for his erstwhile friend. He found him on the corner of Union and C Streets and without warning shot and killed him.

The slayer was arrested on a charge of murder, but at his preliminary examination the charge was reduced to manslaughter, the Court holding that the cruel beating he had received, the shattered condition of his nervous system and the partial derangement of his mental faculties, caused by excessive indulgence in strong drink for so long a period, constituted sufficient proof of his claim that he was in a measure irresponsible for his act. He was held to answer in the District Court for manslaughter, with bonds fixed at $10,000. I will here give an account of Lee's funny, practical jokes while awaiting trial:

One evening in October, 1868, upon the arrival of the stage from Reno, a young fellow dressed in a flowing white robe, stepped out upon the sidewalk and announced to the bystanders that he was the Saviour returned to earth, telling them that God had given him dominion over the world, by making him President of the United States; that he was then on his way to Washington to take over the office and had come among them for the purpose of gathering sufficient funds for the completion of his journey.

him, approached and began talking to him, saying:

"I know you are Jesus, and that it is ordained that you shall rule the world as President. I would gladly furnish you with all the money necessary for your journey to Washington if it was in my power, but I also am poor and have a hard struggle to provide for the needs of myself and those who are dependent on me, but I tell you what I will do. I am running a theater here, and if you will appear on the stage on next Saturday evening, and tell my audience who you are, and what your mission is, I will divide with you, half and half, the receipts of the evening."

These words of Lee rang so true, and he seemed so kind and sympathetic, that the confidence of the young fellow was won and he readily consented to the proposition.

That evening, a "Rush" order was turned in at the *Enterprise* job office for the printing of five hundred of the following posters:

JESUS CHRIST
will make His
First and Only Appearance
at the
Black Crook Theater
Next Saturday Evening
Come one, come all—the biggest time of your life
Don't miss it

These posters were distributed throughout Virginia City and sent by special messenger to all the towns down the line — Gold Hill, Silver City, Empire, Dayton, Carson and Washoe — and by midnight stage to Reno.

When the doors of the Black Crook Theater were thrown open on the evening of the great show the theater soon became packed with a mob of swearing, jostling, wrangling, half drunken roughnecks.

When the curtain rose, and the minstrels had ended their songs and conundrums, Lee stepped to the footlights, and raising his hand for silence, said:

"Friends and fellow citizens, there will now be presented to you, right here on the stage of the Black Crook, the greatest drama of this or any other age. Jesus Christ will now appear before you, and I ask you to maintain perfect quiet during the few moments required to make his entrance."

When Lee stepped aside, the deluded young man, between two girls made up as angels, appeared and walked to the front of the stage. Then Fred Sprung and Johnny Edwards, two husky end men, each with a stuffed club took their places, one on each side of him, and began singing the old hymn, "Give me Jesus." When Fred sang these words, Johnny swung his club, saying:

"Take him, Fred, I don't want him," and gave the poor boy a resounding blow that knocked him reeling over to Fred.

With a mighty blow of his club, Fred sent him back to Johnny, at the same time saying:

"I've got no use for him, Johnny, you can have him."

And so it went. Biff. Bang. From Johnny to Fred, back to Johnny, until at last, the poor, demented, harmless boy went to the floor in a crumpled heap. This brutal spectacle was greeted with stamping, hand clapping, cat calls and shouts of savage joy by the great crowd of ruffians comprising the audience. Cries of "That's the stuff," "Bully for you, Fred," "Keep him a moving, Jimmy," came from all parts of the house. Not a single voice was raised in protest.

When the curtain dropped, the bruised and tortured victim of this funny joke was taken to the hospital unconscious, and remained there two weeks before he could leave his bed.

When Lee was placed on trial for manslaughter, he was acquitted by the jury, and shortly afterwards left the state.

Rivalry Between Express Companies

SECTION 4 of Ordinance 2, prohibiting any person riding or driving through the streets of the city at a rate of speed exceeding six miles an hour, added a considerable sum to the moneys in the City Treasury during the year of 1868, and afforded rare sport to the lovers of horse racing. That year, a sharp rivalry existed between the Wells Fargo and Pacific Union Express Companies, each trying to outdo the other in making time between Virginia City and Reno. This was especially so regarding the letter express. This was carried between the two cities by "Pony Express," the companies having the finest, thoroughbred race horses that could be secured for money.

Upon arrival of the train at Reno, the riders, mounted, would be waiting. The pouches containing the letters would be handed them by the agents of the two companies, and they would start on their race for Virginia City. At the end of each mile, a fresh horse, bridled and saddled, would be waiting. Without dismounting they would spring from the back of one horse to the other and continue at full speed.

Long before they were due to arrive at Virginia City, C Street, from Union to Carson, would be thronged with people waiting to witness the outcome of the race. Thousands of dollars were won and lost on the result of these races, not only by the gambling fraternity, but by the miners, merchants and professional men of Virginia City and Gold Hill. Representatives from newspaperdom were also there, placing their bets.

As the time drew near when the riders were due, the people would become restless and many among them would be seen, watches in hand, craning their necks and gazing in the direction the racers would come. The waiting crowd would at length be thrilled by a shout, "Here they come! Clear the way!"

And here they would come, side by side, neck and neck, horses covered with foam, riders plying whip and spur, until they reached their goal, at the corner of C and Union Streets.

When the horses stopped and the riders dismounted, they were immediately arrested and taken before the City Recorder, who fined them $25 each for violating Section 4 of Ordinance 2. The fine was paid as soon as levied, by the Express Companies.

Law and Liquor

AT the election of 1868, George I. Lamon was elected as one of the Assemblymen from Storey County, upon his pledge to the people that he would introduce a bill in the Assembly for the reduction of the rates of toll on the Geiger Grade. These rates, being so high, were a burden upon the business men of both Virginia City and Gold Hill. He also promised to use all his influence and power in the Legislature to have that bill enacted.

In seeming compliance with his pledge, he did, shortly after the convening of the Legislature, give notice that he would, on the following Thursday, introduce a bill having that object in view; the reporter of the *Enterprise* so stating in his letter of that evening. Thursday came — no bill; Friday and Saturday passed — still no bill. On Sunday night the reporter met him in the Omsby House and said:

"George, what has happened to your bill regarding the rates of toll on the Geiger?"

George looked at the reporter, and with a meaning smile answered:

"Oh, hell! Bill, I have seen the old man and he has made it all right. There won't be any bill."

This man was a fair sample of the majority of the "Honorable Gentlemen" representing the people of the State of Nevada at that session of the Legislature.

It was at this session that all kinds of gambling was licensed by law, thus blacklisting Nevada among the other states of the Union, causing her citizens to be regarded, everywhere, as a Godless, lawless, immoral and degenerate people.

At this time the liquor interest was the supreme power in the state. Every other interest, great or small, was subservient to and controlled by it. No man, from Governor to Constable, could be elected to office who in any way opposed it, and there can not be a doubt that it was through the influence of the liquor interests that this evil gambling bill was placed among the Statutes.

In 1868 there were seventy licensed saloons in Virginia City and fifteen in Gold Hill, making a total of eighty-five for the two towns. Besides these regular saloons every hotel, restaurant and grocery store, and even the confectionery and candy shops, sold liquor by the glass. In every one of the saloons some kind of gambling game was run, such as faro, rondo, roulette, red and black, chuca luck and dice. All these places, especially the faro rooms, were nightly crowded with people, bucking the games.

From 1862 to 1875 there were nearly fifteen thousand miners emplyoed in the underground workings of the Comstock. The daily wage of these men was four dollars per day, giving them an income of $120 per month. Deducting their living expenses, which did not exceed $50, there remained a balance of $70 for them. The greater part of this hard-earned money was passed over the bar for booze or fed into the insatiable maw of the "Tiger."

Consequently, when Virginia City was practically swept from the map by the great fire of 1875, the majority of the miners found themselves without money, food or shelter.

Robert Ferguson, once superintendent of the Gould and Curry Mine, told me that fully ten thousand of them left Virginia City with their blankets on their backs seeking employment elsewhere.

Mark Twain's Boxing Bout

LEAVING crime and debauchery, gambling and booze fighting behind, I will now go, with my readers, into fairer fields and along pleasanter pathways.

While Mark Twain was reporting on the *Enterprise,* the boys got together and raised enough money to rig up a small gymnasium. Mark, being an expert swordsman, was elected fencing master, and Bruce Garvet, a little Frenchman, teacher of the "manly art of self-defense." As Bruce was himself no slouch with the foils, he and Mark gave many exhibitions of their skill with those weapons, all these contests ending with Sam coming out victor.

One evening after Sam had, by a dexterous turn of the wrist, disarmed Bruce, by sending his weapon flying across the room, he stood a few minutes gazing in wondering admiration at Sam, then flourishing his hands, made him a bow, saying:

"Ah! Meester Clemens, you aire one, grande mastair weeth ze foil. I have nevair weetness zo beautiful work like yours and it gives me great plaisure to have play weeth you, but I theenk if you will weeth me spar a few rounds, I weel show you that I am your maister weeth the glove."

Up to this time Sam had always refused to give Bruce satisfaction in a contest with the gloves, saying that it was a brutal sport, and one in which a gentleman should not engage. On this occasion, however, he at last consented to enter the ring for a bout of five rounds. The first two rounds after the men faced each other was occupied in feinting, side stepping and making short jabs to the face and body, neither of them getting in a punch that counted for anything.

In the third round both men began sparring for an opening. Finally Sam, thinking he saw one, rushed and swung for the head, Bruce ducked, stopped the rush, and with a mighty punch to the nose, lifted Sam from his feet and sent him to the mat, where he lay for a few moments, apparently stunned, with a crimson flood pouring from his nose. Finally raising himself to a sitting posture he gazed, first at one of the boys and then at another, with a surprised, bewildered look, which appeared to ask them what had happened. He at last got to his feet, jerked off the gloves, and throwing them at Bruce, made for the lavatory. After washing the blood from his face and getting into his street clothes, Sam, without turning his head this way or that, stalked out of the gymnasium and crossed the street to Morrill's drug store where Dr. Hammond straightened up his nose. He then went to his room and tumbled into bed.

The next morning his nose was swelled to enormous proportions, and his eyes looked like two slits in a black mask. The first friend he met when he appeared on the street was Paddy Keating, who greeted him with

"Jases! Sam, who the divil throwed the brick?"

"Huh!" grunted Sam, and marched on.

The next man was Charley White, who hailed him with, "Great Scott, Sam, what's that hanging to your face? Come in here and I will unhook it."

"Go to the devil!" snarled Sam, and on he went.

And so it was all along the line. Every friend and acquaintance he met having something to say about his nose.

By the time he reached the *Enterprise* building he was boiling with rage, and forgetting all about breakfast, he went on the jump, up the stairs and into the editorial room and without preface, rasped out:

"Joe, I want a month's vacation."

"Oh, you do," said Joe. "Well then, why don't you take it? What have I to say about it? Who are you, anyhow? My friend, I guess you've got into the wrong box."

These words from Joe added fresh fuel to the flame of Sam's wrath and with a mighty blow of his fist upon the table, he almost yelled:

"Look here, Joe Goodman, I've had enough of that gaff this morning from the infernal idiots I've met on the way here from my room. You may think it funny, but I don't see it in any such light and I don't want any more of it. And now, I ask you again for a vacation. I won't stay here to have every damn fool in town shoot off his mouth at me, with some kind of rot concerning my nose, and unless you see fit to grant the vacation I ask for, I will leave without it, and if I do, it will be the end of my connection with you and your paper."

When Sam snapped out his ultimatum, Joe began laughing, but when he saw the frown on Sam's face deepening and the angry glint of his eyes, he became serious again and said, "Oh pshaw, Sam, calm yourself and sit down. Of course, you can have a vacation if you want one. Where will you go, and when do you propose to leave?"

"It is a matter of indifference to me as to where I'll go, so that I get away from here. I will leave on the 4 o'clock stage this afternoon."

"Why, Sam, that's the Austin stage. If you are going out that way, it will not interfere with your work on the paper. On the contrary, it will be profitable to both yourself and the *Enterprise*. During your absence from Virginia City you can not only send a daily letter, but will be able to drum up considerable business for me in the way of ads, bill heads and subscriptions and aside from your regular salary I will pay you a commission on all orders sent in."

Without a good-by to any one but Joe and Steve, Sam left that afternoon for Austin.

Dan McQuill Follows Mark's Nose

WHEN the *Enterprise* came out the morning after Mark Twain's departure, an item appeared among the locals, which set the whole town laughing. This item was headed—"Who Nose Where It Came From" and read something like this:

"Just as the stage was about to leave for Austin yesterday, the people who were waiting to see it off were startled out of their equanimity by a strange and most unusual sight. The passengers were in their places and the driver, Billy Watson, had gathered the lines and was just reaching for his whip, when a prodigious, blood red nose was seen to emerge from the office and start for the coach. When it reached the stage and made an attempt to enter the door, two of the lady passengers began screaming with fright at the sight of the dreadful monster, trying to force its way into the coach. The screams of the ladies caused people to come running from all quarters, up and down the street, to learn the cause of the trouble. An excited crowd soon gathered, and it looked for awhile as though a riot would ensue. At this crisis, however, the agent of the company appeared among them and upon his assurance to the passengers that the Nose was perfectly harmless and gentle, their fears were allayed, the Nose was allowed to enter the stage, Billy cracked his whip, and the stage rolled away.—Continued to-morrow."

Every morning for the next two weeks an item something similar to the above appeared in the local columns of the paper. Dan McQuill trailed Mark's nose along the line of its journey, never mentioning Mark himself — just his nose — giving an account of its enthusiastic reception by the people of all the towns along the way, of the stampeding of bands of cattle and horses, of frightened school children, and its triumphant entrance into Austin, where the Major delivered an address of welcome and tendered it the "Freedom of the City."

These articles almost caused the suspension of all business for an hour or two every morning. People meeting on the sidewalks would stop and start talking, laughing and slapping each other on the back in a perfect riot of merriment.

"Old Dan's a corker, ain't he," would come from one. "You bet he is," another would answer, "but I'm darned if I'd be in his place when Sam gets back. Lord! But wouldn't I like to see Sam now! I'll bet that he's so darned mad that he is dancing a hornpipe all the time. Ha! Ha! Ha!"

Dan's writings, causing so much mirth to his friends, had a very different effect on Sam. He simply went to pieces with anger, and hurried back to Virginia City with a rush. At about nine o'clock on the night of his return, he entered Goodman's room in a state bordering on insanity and blazed out:

"Joe Goodman, whatever in the devil's name possessed you to allow that driveling, half-witted idiot, Dan, to publish his infernal rot about my nose. He has made me the butt of every damn jack monkey from here to Austin. The confounded ass has got softening of the brain, and I will have no further connection with him as a reporter on your paper. If you retain him, I quit, so make your choice."

"Sam," said Joe, "you have worked yourself into a passion which is beyond all reason and common sense. Take a tumble to yourself and think this matter over rationally. Go into the other room and look over the files of the paper for the last six months and you will find that you have made Dan a target for your wit right along. You have stung him pretty hard a good many times, and unmercifully ridiculed him, without a thought regarding his feelings, not caring whether it hurt him or not, just so it put the laugh over on him. Did he ever resent your funny articles at his expense by calling you all the mean things in the dictionary? No, sir! He never whimpered, but took it all in a spirit of fun, and joined in the laugh at his own expense. And now you are climbing all over yourself because he has come back and put the laugh over on you. Cool down, Sam, and forget it. Forgive old Dan, and don't let this trivial matter cause you to break with him. Let things go on in the same good old way, and we will all be the happier for it."

"You term it a trivial matter, do you? Well, I call it a mean and cowardly attack on me, conceived and written in a spirit of jealousy and spite. I will neither forgive nor forget and I again tell you that I will sever my connection with the paper unless you fire Dan McQuill."

"This is a tough proposition you are putting up to me, Sam, and I shall have to take time to consider it. I hope that you will also think it over, and after sleeping on it, change your mind."

After his talk with Joe, Sam went into the reporters' room, sat down and filling his pipe, began smoking. Happening to look up at the wall, he spied some copy Dan had left on the hooks. Taking it down, he resumed his seat and began reading the news item in the familiar handwriting of his old friend, which had been gathered in the earlier part of the evening. He was roused from this occupation by hearing a voice, almost a whisper, saying:

"Sam, may I come in?"

Looking up he saw Dan's good old face, with its long goatee and appealing eyes, framed in the partly opened door. The frown upon his face instantly changed to a smile of welcome, and springing to his feet, he shouted:

"Open that door, Dan, and come in here. You darned old goat, I'm glad to see you."

Then the arms of the two men went around each other, and they began waltzing around the room, knocking over chairs and every other article of furniture that came in their way, bumping against the walls, making such a racket that Steve, fearing that Dan and Sam had gotten into a mix up, hurried to the reporter's room to put a stop to the fracas.

When he opened the door and saw the way Sam and Dan were celebrating their reconciliation, he, too, jumped in and joined in their wild antics. While they were thus whooping things up, Joe came from the other room, and without loss of time, also got into the rumpus. When the jollification ended and quiet was restored, Joe took a check from his pocket and handing it to Sam, said:

"Sam, I hate to part with you, but you have left me no alternative. After considering your ultimatum, from every point of view, I find that I cannot possibly comply with your demands. Dan has been with me from the time of the establishment of the *Enterprise* until now. During that time he has filled his position as reporter on the paper so ably and has been such a true and faithful companion and friend that it would be an act of gross injustice, not only to him, but to the whole community as well. Here is a check for your salary, with an additional fifty dollars as a token of my appreciation of your past services."

When Joe presented the check, Sam rasped out:

"Don't make a fool of yourself, Joe. Throw that confounded check in the stove, and go back into your room and see if you can, for once in your life, write something with a little common sense to it. Come on, Dan, let's go down town."

And these two funny men walked out, arm in arm.

Mark Throws Up His Hands

TWO nights before Sam gave his lecture on the Sandwich Islands, at Gold Hill, three men had been held up and robbed on the "Divide" by a single bandit.

The next day while talking the occurrence over with Steve Gillis, Sam criticized the three victims of the holdup in pretty strong terms, saying,

"Steve, I can't even imagine how three big, husky men could tamely submit to being held up and robbed by one man. Great Scott! I should think they'd go and hide themselves for very shame, the darn cowards."

"These men may not be cowards at all, Sam," said Steve. "These holdup men never stop a man unless they know they have the drop on him. Every one of these hounds is an expert gunman, too, and, in my opinion, it is no evidence of cowardice for a man to submit to their demands, when he knows that he is taking a big chance of losing his life by resisting them. Put it up to yourself, Sam. Suppose, for instance, that you and a couple of companions were on some cold, dark night, with overcoats buttoned to the chin and hands in your pockets, coming up the 'Divide' to Virginia City. What would you do if you suddenly found yourself looking into the muzzle of a big 44 Colt and heard a rough voice commanding you to throw up your hands. Would you do so, or fight?"

"While I don't suppose I will ever be called upon to act in a contingency like that, you can just bet your life, Steve, I wouldn't crawl," said Sam.

These bold words at once determined Steve to play a big practical joke on Sam, by stopping and robbing him the night of his lecture at Gold Hill. As Dennis McCarthy was Sam's agent, he, of course, had to be taken into the plot. Dennis, appreciating the fun of the thing, at once joined in the conspiracy.

Steve then hunted up Joe Harlowe, Little Hicks, Salty Boardman and John Russell, and they, too, became members of the Gang. Joe Harlowe and Hicks were to do the robbing, while Boardman and Russell were to remain in the shade so that Sam could just distinguish them through the darkness. Steve was to wait in the composing room to receive Sam after the holdup.

When the curtain fell on the night of the lecture, Sam and Dennis, after depositing most of the money taken in at the box office with Wells, Fargo & Co., to be forwarded to Virginia City, started on their way homeward. They had gotten to a point nearly opposite the old shaft of the Imperial and Empire, when "Hands up!" came a sharp command from right beside them. Sam didn't crawl, but his hands went into the air with promptness and dispatch.

Then, "Bill, you keep these two nuts covered while I go through them," came a rough, grating voice. "And you fellers keep your hands up, and don't try any monkey business. If you do, it will be the worse for yer."

After relieving Sam and Dennis of their money, the robber who had searched them said, "Now you fellers listen. If you move one step from this spot in less than half an hour, you will both be dead men."

So saying, both robbers backed away, and disappeared.

The night was a bitter cold one. A strong north wind was blowing and the air was filled with drifting snow. Dennis had no overcoat, and had become almost half frozen while waiting for the robbers to finish their job. As soon as they left, therefore, he chattered, "Come on, Sam, let's get out of this."

"Dennis," said Sam, who had on a heavy overcoat, "that big duffer told us that if we left in less than half an hour we would be killed. As I have no desire to leave the world at the present time, I am going to stay here until that half hour is up."

"Why, Sam, do you suppose that they are going to wait to see if we leave or not. Both of them are at the top of the hill by this time, beating it for town. Come on and get out of this wind and snow. I'm near frozen."

"You may be right, Dennis, and you may be wrong, but I am taking no chances either way, and I am not going to move from this spot till I know that I am taking no risk in doing so, and you are going to stay with me."

When the victims of the holdup reached the *Enterprise,* they went into the composing room to warm themselves by the big box stove, in which there was a rousing fire. The other boys, Dan McQuill among them, were there waiting. As they drew near the stove, Steve greeted them with:

"It seems to me that you two fellows have been a mighty long time getting here from Gold Hill.

What's been keeping you; been stopping by the roadside to rest?"

"We did stop once by the wayside," answered Sam, "but it wasn't to rest by a long shot. We met two gentlemen on the 'Divide' who thought they needed what money we had more than we did, and they argued their claim so convincingly that we let them take it. The long and short of it, boys, is that we've been held up and robbed."

"By gracious, is that so, Sam?" said Salty. "Did they get the money taken in at the box office?"

"No, a few dollars only. The receipts are in the safe of Wells Fargo. But I tell you, boys, Dennis and I had a mighty close call and it looked pretty nasty for a while, with two big guns within six inches of our eyes."

"How many were there, Sam?" asked another of the boys.

"I don't know how many were in the gang. I saw four, though only two of them had their guns trained on us. But those two big 44's had a mighty persuasive look and were just as effective as a whole battery of cannon, but 'All's well that ends well,' and I am glad that I am back with you fellows instead of at the bottom of one of the old holes out there on the 'Divide.' "

"So am I, Sam," said Steve, "and I'm awful glad that you didn't crawl."

For the next two or three days Sam was kept busy recounting his experience with the robbers. At the end of that time he was made wise to the joke by Salty Boardman, the robber chief halting him on the street and ordering, "Hands up!" in the gruff voice of the robber, at the same time pointing a dummy gun at him and then sprinting upstairs into the composing room.

A few minutes previous Sam had noticed Salty and Dennis talking and laughing together and as he approached them, Dennis, still laughing, walked away, so that when Salty repeated the words, "Hands up!" and ran up the stairs, he at once knew that Dennis was in on the holdup.

If Sam had ever been mad in his life, he was mad now. He was simply speechless with fury. Going into the publication office of the *Enterprise,* he wrote a check, and started on a hunt for Dennis. He found him at Charley Legget's eating house. Throwing the check on the table, he said, "I do not require any further service from you, sir," then turned and walked away.

Going back to the office, he walked up to Dan McQuill and asked, "Did you know that this confounded fake holdup was going to be put over on me?'"

"No, I did not. I knew no more about it than you did," replied Dan.

"Do you think Steve had a hand in it?"

"Well, you know that Steve is always wise to any fun among the boys, but I know that he was not one of the robbers because he was at work in the composing room from five o'clock until you got back from Gold Hill."

"I hate to think that Steve would go back on me like that, but I'll bet he was in it up to his eyes. And you bet I'll get even with him."

From this time until he left Virginia City, Sam would have nothing to do with any of the boys, refusing even to speak to them, Steve among the rest. In the course of a conversation had with him, shortly after his return to San Francisco, he said to me:

"Billy, if Steve had given me an inkling of that holdup, I could have turned the tables on them, and they would have been caught in their own trap, and the big joke would have been on them instead of me."

"Yes, Sam, it would, but that would have spoiled everything. The plot was to rob Mark Twain. In consequence of that robbery, Mark Twain has gotten a good many thousand dollars worth of free advertising."

"Oh, you are just as bad as the rest. Stand in with a pack of coyotes against me, of course."

"No, I am not standing in with any one against you. I am just telling you, as a friend, what I think about it. Sam, you know every one of the boys is your friend. You know, also, that you could not ask any favor or service of any one of them that they would not grant or perform. I am going back to Virginia City in a few days. Let me be the bearer of a message of reconciliation from you and we will all be the happier for it."

"Oh, well, Billy, do as you please. I really have nothing against the boys, only that they always single me out to play their jokes on. Why don't they go after somebody else once in a while and leave me out?"

"It is because it is you, Sam. What fun would it be to hold up me, for instance, compared with you? I am nobody and the thing would fall flat and be no joke at all. But when you, Mark Twain, known all over the country as a traveler, writer and lecturer, have a joke like this played on you by your friends, it is a real, big, practical joke; a joke which helps you to make the people laugh wherever and whenever you appear before them."

Sam then shook hands with me, and as he started to walk away, turned and said: "You can tell the boys that you saw me, anyway."

I Go into a Funk

THE next time that I saw Sam was in Virginia City at the time of his lectures on the "Holy Land." I was then associated with Dan McQuill as a reporter on the *Enterprise*. I had been assigned to report the hanging of John Milaine, the first man executed, by law, in the state.

It happened that the hanging was to be on the day of Sam's first lecture. My very soul revolted at the very idea of witnessing the gruesome sight, let alone reporting it, so that when Sam came into my room on the preceding evening, I said to him:

"Sam, I am going to ask you to help me out of the biggest and deepest hole I was ever in. I'm clear down at the bottom, Sam, and unless you pull me out, I'm gone."

"What do you mean, by being in a hole? And how am I going to pull you out? What's the trouble?"

"The trouble is that I have been assigned to report the hanging of John Milaine to-morrow and it is so horribly repugnant to me that I would almost as soon die as to be present. So I'm asking you, Sam, to show your friendship for me by taking my place."

"Oh, that's it, is it? When Joe put you on here with Dan did you think your experiences as a reporter would be altogether pleasant ones; reporting picnics, dances, social gatherings, political meetings and all that, with a free banquet now and then? Did the roads all seem nice and smooth, with no check holes and rocks in them? If that is what you expected you've fooled yourself. And if you're going to funk when you run into a tough proposition like this, you will never amount to anything in newspaper work and you had better go back to mining at Jackass Hill. Come, buck up, Billy. Put your own feelings aside and remember only that you are there to write an account of the hanging for the paper, and you will pull through all right."

After this lecture from Sam I did buck up in a way and went out to the place of execution the next morning determined to hold myself together the best I could. When I got there the hills surrounding the little valley in which stood the scaffold were covered with thousands of people. Men, women and children, many of the women with babies in their arms, had gathered to witness the gruesome spectacle.

Aside from a feeling of disgust at the morbid curiosity exhibited by these people, I got along fairly well. But when the condemned man mounted the scaffold and took his place on a trap and Sheriff Cummings assigned me a place among the other reporters, directly behind him, I came very near losing my nerve. But I pulled myself together and stood there, while he was making his farewell speech to the world and its people, with a pretty good hold upon myself.

After thanking the sheriff, and all those who had been kind to him during the time of his confinement, he turned toward the crowd, and making an elaborate bow, spoke for about half an hour in a rambling, disconnected sort of way, ending with:

"I am a Frenchman, but you must not think you hang all France, when you hang me. There are plenty of Frenchmen left."

While the sheriff was adjusting the noose, a feeling of horror again came over me, and when the trap was sprung and he shot down to his death, I simply went into a blue funk, and getting off the scaffold, I went back to Virginia City as fast as my legs would carry me.

Mark Lectures on the "Holy Land"

SAM lectured two evenings on the "Holy Land," while in Virginia City. The first of these lectures I did not attend, but was present at the second. I wish that I could fittingly write of that lecture and of Sam himself. It would be futile for me to attempt to do so, but I will say that of all the men I have heard or seen upon the lecture stage, he was the greatest and most magnetic.

His person, his poise, his manner of delivery and his drawling voice, were so perfectly in accord with his subject that from the first word to the last, the interest of his audience never flagged. He would at times have the people in tears by telling them stories of suffering in the Holy City; of little children starving in the streets; of decrepit old men and women sitting by the wayside with shriveled arms and hands held out for alms; and of the lame, the halt and the blind, huddled together to warm their diseased and almost naked bodies.

He would then throw them into convulsions of laughter with one of his ludicrous funny stories, all the while looking as solemn as an owl.

He described the scenery along the banks of the Jordan in a captivating way, and then took them with him along the Jericho Road. His tribute to the Good Samaritan was paid in most beautiful language, so full of pathos and tenderness, that he took the hearts of his hearers by storm, and when he paused, the audience arose and greeted him with a cheer that shook the theater.

His description of a full-rigged clipper ship, homeward bound, under all sail, was the most beautiful word picture to which I ever listened.

He told how the passengers on his own ship went wild with joy when they saw their country's flag go to the peak in salute; how they embraced each other and cheered and laughed and cried. He ended by telling them how the captain's son sprang to the rail and recited the last stanza of Drake's address:

> "Flag of the free heart's hope and home,
> By angel hands to valor given;
> Thy stars have lit the welkin dome,
> And all thy hues were born in Heaven.
> Forever float that standard sheet,
> Where breathes the foe, but falls before us;
> With freedom's soil beneath our feet,
> And Freedom's Banner floating o'er us."

When Sam ended his lecture with these lines, his audience went wild, and he stood bowing and smiling for full ten minutes before the cheering and other demonstrations of appreciation ended.

He lectured at Gold Hill the next night and at Carson on the one following, being received with the same cordiality and enthusiasm as at Virginia City. His tour came to an end with his lecture at Carson. From there he returned to Virginia City and, after bidding his friends good-by, left the state.

I never met Sam after this parting, but the personality of this wonderful man is indelibly stamped upon my memory and he stands just as clearly before me to-day as in the years of Long Ago, when we were both young men in the heyday of our lives.

Spiritualistic Manifestations

ONE evening, when in Gold Hill, I walked in on Alf Doten at the news office and found him engaged in whittling out something from the lid of a cigar box.

"Hello, Alf," said I. "What are you making?"

"I am trying to make a Planchet, but I don't know whether I am going to succeed or not. This blamed wood splits so easy that it's hard to shape it with a pocket knife and I don't know but that it would be cheaper to send to the city and get one."

"What are you going to do with it when you get it, Alf?"

"What am I going to do with it? Why, I am going to use it to communicate with our Spirit friends, of course. Didn't you ever see a Planchet work?"

"Oh, yes, I've seen my sisters and our friends having lots of fun by shoving one of them over a sheet of paper and writing foolish answers to foolish questions, but the idea of communicating with the Spirit World never entered my head. I always thought they were a kind of fortune telling contraption, just to have fun with."

"Oh, you are way back in the dark," said Alf. "See here, Billy, about a half dozen of us get together every Saturday night in Billy Van Bokkelin's room at the Chollar to investigate along the lines of spirit communication. Get into the ring, and you will both see and hear things that will open your eyes to the truth, that our friends of the other world do come back to us."

"All right, Alf, I'll be present at your next meeting, but I won't promise to become one of your members."

When I got to the place of meeting at the appointed time I found the members of the "Circle" already there, with hands on a table. When I entered, Alf motioned me to a chair opposite himself. Seating myself, I also placed my hands on the table. We had sat there in silence for perhaps five minutes, when without a word being spoken or a question asked, there came three distinct raps on the table.

"Who are you?" asked Dagget.

When Dagget asked this question, Alf, who was the medium, fell back in his chair with the muscles of his face twitching in a curious way, then his head began to jerk from side to side and his whole body seemed convulsed with a heavy chill.

"Well, speak up," said Dagget. "Who are you, and what do you want?"

"Ugh! Heap big Injun," came in deep guttural tones from Alf. "Roller no good, heap make sick, Injun die."

"Well, I'll be darned," said Putnam, "if it ain't old Piute John who poisoned himself with the printer's ink that was on the press roller?"

"You've told us that before, John. Is there anything else you want?"

"Injun want Squaw Jim catchum Injun's squaw. Heap killum Jim, damn Washoe Injun."

"All right, John. We'll fix him. You now go wooksum."

Before Alf came out of his trance, there was a rap on the table that sounded like it had been made with a sledge hammer, and it began tipping and hopping all over the room, finally losing a leg and falling to the floor with a crash, breaking the lamp and smashing a chair in its fall. This was the spirit of Eugene Aram, who was hanged in England many years ago for murder. He was a man of great learning, and led the life of a recluse, being almost entirely absorbed in his studies.

His gentle and retiring disposition and his readiness to do an act of kindness won for him the highest esteem of all with whom he came in contact so that, when the charge of murder was made against him, it seemed so at variance with his past life, that the people who knew him could not bring themselves to believe in his guilt.

When he was brought to trial, however, circumstances pointed so directly to him that although the doubt of his guilt was not removed even from the judge who sentenced him, he was convicted and hung. From that day to this that doubt has continued to exist in the minds of thousands of people all over the world. Ever since then Eugene Aram has been trying, through the medium of Spiritualism, to convince the world of his innocence.

On this night when this seance with this spirit had ended and he had taken his departure in the same violent manner of his entrance, Billy Van Bokkelin snapped:

"Now you fellows just listen to me a minute if you please. Every time that crazy gink comes in here he gets mad and goes to smashing things. It has already cost me more than fifty dollars to replace the things he has broken. Not one of you have ever offered to put in a cent to repair the damage. I am not going to stand it any longer, and unless you bar him in the future, you will have to find another meeting place and leave me out."

I attended several of these meetings afterwards, but never joined the "Circle." It was about the same old thing over and over: rappings, table tippings and Planchet nonsense, with Alf's usual contortions, while under the control of "Heap big Injun," a "Gypsy Queen" or the old prospector who had been killed by the Indians just after having discovered a fabulously rich gold mine. I had witnessed all this kind of doings before. They soon palled on me and I quit going.

Billy Van Bokkelin was Secretary of the Chollar Potosi Mine, and spent much of his time experimenting with high explosives. He had a little monkey named Cap. Cap was always in evidence when Billy was in or around his office, trying to imitate anything he saw his master doing.

One day while Billy was in his office, the whole building was utterly destroyed by an explosion, in which he lost his life. The cause of the accident was never known, but it was generally thought that it occurred while he was experimenting with the explosives.

I had left Virginia City prior to this time, but my old friend, Putnam, who was on a visit to me in 1886, gave me a full account of it, telling me at the same time of a seance at which Billy manifested his presence and told them all about it.

"We had been talking about Billy and speculating as to the cause of the explosion, when there came three quick raps on the table. Alf immediately stiffened and went into a clairvoyant state. Then we heard a voice, which we all at once recognized as that of Billy Van Bokkelin."

"Hello, is that you, Billy?" asked Ed Feusier.

"Yes, it's me all right," came the answer.

"We were talking about you when you rapped, and were wondering as to what caused that awful accident. Can you tell us, Billy?"

"Oh, yes! I was in the laboratory making some tests. There was a big box of giant caps under the table where I was at work and that little rascal of a monkey in some way got hold of one of the small boxes and was trying to pick the fulminate out of one of the caps with a nail. It went off, of course, and — well you know what happened then."

"It must have been a terrible shock to you, Billy. Did you realize what happened?"

"No. I felt no shock whatever, and only realized that I was going, at lightning speed through utter darkness. I had been traveling in this way, it seemed to me, for ages, when suddenly the whole space lit up with a dazzling brightness and I saw that I was approaching a great lake of surpassing beauty. The shores of this lake were covered with lovely flowers of every shade and hue imaginable, while magnificent trees, with branches drooping over the water and their silvery foliage shimmering in the brilliant light surrounded the lake on every side and myriads of fish having all the colors of the rainbow disported themselves in its crystal clear waters. From the great shining city built on the gently sloping hillsides, a broad white highway led down to the shore. On this highway thousands of radiantly clad people were walking up and down, to and from the city, while thousands were sitting or reclining under the trees in friendly converse. Still others were gathering shells and shining pebbles along the shore and an atmosphere of love seemed to pervade everywhere. I had been standing in this beautiful country but a short time when I was approached by a spirit robed in white, with a golden band circling his brow, and a silver star set in the center of his forehead, who said:

"'Come with me. I must now conduct you to the palace of the King, where you will be judged as to your fitness to remain with us.' He then took me by the hand and led me along the highway to a large building, which seemed to be of pearl.

"After entering the door I was conducted into a chamber glittering with all manner of precious stones. There was no artificial light in this chamber, neither was there sunlight, but the whole place was refulgent with a soft, mellow light emanating from a benign Being sitting on a crystal throne at the far end of the chamber. There was nothing fearful nor even austere about this glorious Spirit. His face shown with Divine Love and Pity. When I stood before Him at the foot of the throne, He asked, in a gentle voice:

"'Whence come you?'

"When I answered, 'Virginia City,' a note of sadness was in His voice as He said:

"'Very few from that city are found worthy to enter here.'

"Witnesses were then called upon to testify as to my conduct while I lived on earth. The preponderance of the evidence was very much against me and I surely thought I would be cast out, but just as I had almost given up hope, the spirit who had conducted me to the Judgment Hall appeared at my side and told how I had taken that little darky, who had both legs broken by an express wagon running over him, to my home and cared for him until he got well, and then set him up in the bootblack business. At the conclusion of this spirit's testimony, the King said to me:

"'Inasmuch as you have done it unto one of the least of these my Brethren, you have done it unto me.'

"I was then told to go and join the other spirits. I tell you, boys, that in the end it pays mighty big to do an act of kindness, however small it may be."

"Have you met any one you knew in this life, Billy?"

"Oh, yes. A great many of them."

"Do they look anything like they did here?"

"In one respect, yes. That is, I know them as soon as I see them, but in all other respects they are entirely different. Their earthly bodies have been changed to spirit bodies; young, perfect and beautiful, and their eyes are luminous with joy and love."

"Have you seen any of the wise and great, who once lived here on earth?"

"I have met with the spirits of all the good ones. The cruel and wicked ones, such as Nero, Caligula, Queen Catherine, and all the other bloodthirsty despots, are barred from entrance here."

"Do the spirits of the great mingle with the other spirits or do they hold themselves aloof?"

"There are no distinctions between them here. There is no one spirit greater than another and they are all dwelling together in perfect harmony, peace and love. Boys, this is Heaven."

With these words our seance ended. Alf came out of his trance and Billy left us.

I had listened to Put's story with intense interest. When he finished he lit his pipe and looked over at me with a twinkle in his eyes, that plainly asked: "Do you believe all this?"

I shook my head at him and said: "You are an old humbug. Come out in the garden, where you can fill up on cool air."

Dear old Put. His heart was a heart of pure gold; loving everybody; everybody loving him. The world was made better because of his citizenship and he was a blessing and benediction to all who knew him. It was he who presented the bogus pipe to Mark Twain.

Burkhead, to whom the genuine pipe was given,

had been very popular with all the boys in the office except one, while foreman of the composing room. Jimmie Sweet was a good printer and usually set a clean proof, but he got into the habit of going to a faro game after he was through with his work and playing the game until daylight. Consequently he would frequently foul his case by throwing in his letter while half asleep, and, of course, his proof would be a very dirty one. Burk would go after him red hot, calling him a blacksmith, a slouch and a noodlepate and threatened to fire him if he didn't wake up and do better. He sometimes told him that he was a disgrace to the office and to the "Craft" and that he had better get a job on a hay ranch.

Sweet never resented these scoldings, but would acknowledge his fault and promise to reform, always ending with "You are right, Burk. You're always right. And Burk, old boy, you're a diamond to keep me on."

On the night of the presentation, after all the boys had told Burk good-by, Sweet went up, and extending his hand, said:

"Well, Burk, good-by and good luck, old man. I've always said you were a diamond, and I say so now. By George, you are more like a diamond than any man I ever saw."

Burk was highly pleased at these, apparently complimentary and friendly words, and became all smiles as he grasped Sweet's hand and said:

"Thank you, thank you, Jim. But tell me in what way do I resemble a diamond?"

"Oh, because you are a seven-sided old son of a gun," replied Sweet. Then he skipped.

Mark Twain's Fake Duel With Sam Leonard

I READ some months ago, a most interesting article written for the *Stockton Record* by a former compositor on the Virginia City *Chronicle,* giving an account of a trip made by him to Nevada. In this article he mentioned the famous duel between Mark Twain and Sam Leonard, stating that it never occurred. He was mistaken. The duel did take place and was fought to a finish, although there were no casualties.

Leonard was a compositor on the *Enterprise* and was well liked by every one connected with the paper. But he at times became tiresome in boasting of his lineage as a Kentucky gentleman of the "Old School." Mark Twain cared little about "family trees." One night while Leonard was thus boasting about the Generals, Admirals and other distinguished men among his ancestors, telling of their high attainments and their great wealth in slaves, cotton and tobacco, Mark interrupted with:

"Sam, the tobacco growers often met with serious losses by the ravages of the tobacco worm, did they not?"

"You are right, sir, they did. I once saw a field of a thousand acres, belonging to my father, almost entirely destroyed by them."

"Was there no way of preventing this havoc?" asked Mark.

"Only one, sir. only one. They had to be picked off by hand, sir."

"The worms picked off a thousand acres of tobacco by hand? Why, Sam, it seems to me impossible."

"It may look that way to you, sir, but when four hundred niggers get to work in a tobacco patch, it don't take them very long to clear it of the worms. The great trouble is that they have to go over and over their work day after day until the plants are cut."

"Well, that's a job that wouldn't suit me," said Mark Twain. "Didn't it make your hands nasty, Sam? I don't see how a man like you ever endured such filthy work."

"I will give you to understand, Mr. Clemens, that I was never employed in such menial tasks. And I will have you know, sir, that your remarks are highly offensive and grossly insulting. You will hear from me in due time, sir," said Sam, as he left the room in high dudgeon.

The next morning while Mark was seated at his desk, engaged in writing up his notes, Joe Harlowe entered the room and handed him a note from Leonard. It read like this:

"To Samuel L. Clemens, otherwise known as Mark Twain. Sir: My feelings and dignity as a gentleman and a man of honor never sustained so great a shock as that caused by your unwarranted and brutal attack of last night. I cannot condone or overlook your outrageously insulting remarks. I therefore demand from you a full and ample apology. If you refuse to make this apology, I demand that you give me the satisfaction of a gentleman, usual among men of honor. If you fail to comply with my reasonable demands, I will publish you as a poltroon and a coward.

"This will be handed to you by my friend, Mr. Joseph Harlowe, who is fully authorized to act for me in making all preparations in the premises, with any friend of your own to whom you may see proper to refer him.

"Yours, etc.,
"SAMUEL LEONARD."

When Mark Twain finished reading Sam's "Defi," he looked up and said:

"Joe, is this a real, genuine challenge from that swelled-headed old wind bag?"

"Yes," replied Joe, "it is. But I have talked it over with Steve, little Hicks and some of the other boys and if you will fall in with our plans we'll put the biggest joke of his life over on Sam."

"Just tell me what kind of a fool scheme you fellows have got into your heads now. I can't see for my part how you are going to work a joke into a duel."

"Well, you will accept the challenge and name pistols as the weapons. Then Steve and I will load the guns with powder only. The principals will then take their places at a distance of ten paces. At the word they will fire, take one step and continue firing until the combat ends. Sam, knowing nothing of the manipulation of the pistols, will, of course, believe that everything is being carried out in the regular way and after the first fire we feel sure that his honor as a Kentucky gentleman will have been fully vindicated. Then it will be shake hands all around and we'll all come back as good friends as ever."

"How many of you 'Trick Ponies' are in on this game, Joe?" asked Mark.

"Well, there's Steve, Hicks, Dan and Put, besides you and me. And in order to impress Sam with the idea that he is going to face you in a real duel it will be necessary to have a doctor present, so we have let Dr. Hoyt into the secret.

"Oh, well, go ahead. But look here, Joe, be sure that you and Steve make no mistakes in loading the pistols."

Two mornings afterwards, in accordance with this program, Mark and Sam, with their seconds and the witnesses, arrived at sunrise in the little valley where the duel was to be fought. When the two men took their positions, after the usual preliminaries, Put stepped to the front and asked, "Gentlemen, are you ready?"

Upon both men assenting, he said:

"I will now count one — two — three — fire. At the word fire you will raise your weapons and fire; then advance one step and fire again; and so continue to advance and fire as long as you both remain standing."

When he announced the word "Fire," both guns were discharged almost simultaneously, and almost instantly afterwards Sam was seen to drop his pistol, and after wavering a moment, fall to the ground. At this serious and unexpected outcome of their joke, the boys, thinking that Steve had in some way made the mistake of ramming a bullet instead of a wad into Mark's gun, at once gathered around him with consternation written on every face. Mark Twain knelt at his side and taking his hand, said in a quavering voice:

"Sam, are you really hurt? My God! I had no intention of hitting you and I would not have had this happen for all the money in the world. Forgive me, Sam, won't you?"

Sam looked up at Mark with a sickly smile on his face and said, in a feeble voice,

"It's — the — fortune — of — war, sir, the — fortune — of — war. You are — the victor. I — am — satisfied."

At this juncture Doctor Hoyt, with his instruments in hand, shoved the boys aside and stooped to examine the wounded man. When he unbuttoned and opened Sam's vest, a paper wad was found lodged between the vest and his shirt, directly over the heart. This wad was what had struck Sam and when he felt its impact on his breast he thought that he must be mortally wounded. When little Hicks saw that wad, he snatched it up and put it in his pocket so that Sam would never know what hit him.

When Doctor Hoyt reported that his patient

was not seriously wounded, but was suffering from the shock more than anything else, the boys found they had been more scared than Sam had been hurt. Their joke had acted like a boomerang and had rebounded on themselves. When Sam got to his feet they all went back to Virginia City with faces as solemn as the chief mourners at a funeral.

Dr. Hoyt was the Dean of the *Enterprise* Chapter, of the Virginia City Typographical Union. He was a man who commanded the respect and confidence of every one with whom he came in contact. Among the printers there was no one like George, as they affectionately called him. He was counselor, comrade and friend to them, always ready to help or to do an act of kindness.

At the time of this duel he was about seventy; old in years, but in every other way a big, merry-hearted, fun-loving boy. The doctor and I were very close friends and he often came to my room and spent an evening with me, recounting his experiences and impressions while traveling around the world; describing far countries and strange cities, together with the customs, religions and habits of their people. He would also describe to me the different animals, birds, reptiles and fishes; also the fruits, flowers, trees, shrubs and other growths. He would take me into trackless wastes of burning sand and over mountains covered with snow; then seeing that I was oblivious to all my surroundings, he would look at me with a quizzical smile and bring me back by saying, "Well, what do you think of it?"

In the winter of 1867 an epidemic of confluent smallpox swept over the greater portion of western Nevada, being especially severe in the towns along the Comstock. Very many deaths occurred from the ravages of this dread disease in all these places. Consequently a panic ensued among the people and hundreds of them fled from the state in fear of their lives.

When the disease was at its very worst in Virginia City and the pest house was overcrowded with its victims, Dr. Hoyt, who was a physician and surgeon as well as a printer, threw up his case on the *Enterprise* and joined the other doctors in their endeavors to quell it. He took hardly any time for rest or sleep, but was continually on the go, day and night, visiting and treating the sick; taking no thought of himself. Being an old and not too vigorous man, his strenuous work at last told upon him and he was stricken by the loathsome disease.

When on his way to the pest house, where he went when he found he had the smallpox, Joe Harlowe and I saw him walking up the center of C

Street and started to intercept him. When he saw us approaching he stopped, and shaking his fist at us, yelled:

"Go back, you young fools. I've got the smallpox. Don't come near me."

He then took from his coat pocket a small leather case, which he placed upon the ground, saying:

"You will find my watch and chain in this case. I am going to the pest house. If I die I want one of you boys to have them. You can decide as to which one of you shall own them by either pitching heads or tails or drawing straws, but don't play cards or throw dice for them."

He then told us good-by and went on his way. The doctor did not die, however, but came out after two weeks' internment, cured and with no appearance of having had the smallpox except one small pit on the end of his nose.

I Put a Hot Pitch Plaster on Doctor's Back

IN January of the following year, 1868, Dr. Hoyt had a severe attack of rheumatism in the small of his back. This at times made him crabbed and peevish and if at such times his "fur was rubbed the wrong way," the fellow who did the rubbing would have to look out for himself for the doctor would fly at him like a wild cat and he would be a lucky man if he escaped without being scratched.

I once unintentionally got myself "in bad" with him for a while by that kind of rubbing. I had been all over Chinatown gathering all the news items I could pick up among the Celestials regarding the Tong War, then going on. I had met with little or no success for, as every newspaper reporter knows, it is like drawing blood out of a turnip to get anything out of a Chinaman that he doesn't want to tell. It was "Me no sabe" this and "Me no sabe" that, in answer to almost every question asked, so that when I started home just before dark I was about as wise

I PUT A PITCH PLASTER ON DOCTOR'S BACK

regarding the object of my trip to Chinatown as I was before I went.

When I got to the *Enterprise* building I was chilled through and through, so instead of going up the stairs, I went down into the press room to thaw out. I had been there talking to Andy Peasley for perhaps half an hour when Dr. Hoyt came down the steps with a pitch plaster in his hand.

Coming to the boiler where I was standing and handing it to me, he said, "Billy, I want to get this plaster good and hot and put it on my back. Joe Harlowe put one on me last night but the darn ninny didn't get it hot enough and it wouldn't stick. I want you to heat this one, so that it will stay on my back."

"All right, Doctor, I'll fix it for you."

I then held the plaster close to the boiler until it seemed hot enough and said, "Doctor, I guess the plaster is all right now. Hold up your shirt and I will put it on for you."

The doctor turned a scowling face to me and snarled, "Darn it. Do what I tell you. Get — that — plaster — hot."

"All right. I'll get it hot enough this time." And I again held the plaster to the boiler, keeping it there until the pitch began to burn my hands.

"Look here, Doctor Hoyt, I can't hold this thing

any longer. It's as hot as blazes and if it goes on your back now it is going to make you bounce, old man. You'd better let it cool off a little."

"It's my back, confound you. It's not going to burn you. Slap it on!"

"Very well, then slap it on goes, but I'll tell you again, you'll wish I hadn't. Stoop over and — here, Andy, you hold up his shirt while I put the thing on — but I warn you once more . . ."

"Shut up, you fool," yelled the doctor, "slap it on."

And on it went. When the boiling pitch came in contact with his back, the doctor went up like a Jack-in-a-box, his head striking Andy under the chin, causing him to nearly bite off the end of his tongue.

"Yeow! Gol darn you! Take it off! Take it off!" he howled. But the doctor was performing so many wild stunts, jumping up and down, first on one foot, then on the other, whirling round and round and trying to get at the plaster with his own hands, all the while letting out frightful war whoops, that neither Andy nor I could get hold of the plaster. Finally my poor old friend bounced out of the press room and up to his own room, where he locked himself in.

For the next week the doctor had no use for me. He would pass me by as though I was an utter stranger, and whenever I spoke to him, would scowl at me and tell me to go to the devil. I was, therefore, greatly surprised one morning when a few minutes after having greeted me in this way, he came into my room and offering me his hand, said:

"Billy, I've been acting mighty mean to you for the past week, and I am really ashamed of myself and have come to ask you to overlook and forget it. But I tell you that infernal plaster was the hottest damn thing that ever got a hold on me. It not only burnt where you stuck it on but my whole body from my head to my feet felt like I had been dipped in a barrel of boiling tar."

"Well, Doctor, you know that I warned you beforehand that it was altogether too hot, but you would not listen to me."

"Yes, I know, but my back was hurting me so badly at the time that I never thought of the plaster, only that I wanted it hot enough to stick and you bet I got my wish. But I have not had a twinge of rheumatism since and I believe that I am cured, but there will never be another pitch plaster on my back."

Mark Twain Becomes Private Secretary to Senator Stewart

WHEN Wm. M. Stewart returned to Virginia City after he was elected U. S. Senator by the Legislature of Nevada, he entered the reporters' room one evening, where he found Mark Twain engaged in writing up his items for the next morning's issue of the paper.

"Hello, Sam," said he. "Busy at work, I see."

"Well, yes, Bill, that is, I have been, but I am nearly through now. Take a seat, light up your pipe and make yourself comfortable and I will be at your service in a very few minutes."

After finishing his writing, Sam hung his copy on the hook and, turning to Stewart, said:

"I suppose it will not be long now when you will get the scent of the sage brush out of your nostrils, shake the sand of Nevada from your feet and join the other 'Wise Men' of the nation at Washington. By George, Bill, I wish I was going with you."

"You do!" exclaimed the senator. "If you mean that, Sam, your wish shall be gratified. That is my reason for calling upon you this evening. You are the one man of all others whom I should like to have as my private secretary and if you will accept the post, it will not only deeply gratify me, but will be of great advantage to you. It will bring you in touch with some of the greatest minds of the nation, and give you the entrée into the very best society in Washington. It will also give you a better understanding of governmental workings, both in this country and in Europe. In fact, Sam, it will be beneficial to you in every way and I sincerely hope that you will accede to my wishes."

"Bill, I appreciate your kind and flattering offer to the fullest extent and I most heartily thank you for it, but I am afraid that your partiality has caused you to overestimate my ability to fill a position of this kind satisfactorily and creditably to myself. I think, therefore, that we had better, as Judge Rising would say when not sure of his ground, take it under advisement for a while. To 'act in haste may be to repent at leisure.' I have had one experience of this kind and don't want another. When my brother Orrin was elected Territorial Secretary he appointed me as his private secretary. Well, I held that job down two months. Then he fired me."

"Your brother fired you, Sam. What for?"

"It was simply for answering a letter from one of his correspondents in a way that didn't suit him. He was in Virginia City at the time and expected to be away for two weeks. The day before he left, he said to me, 'Sam, in looking through my correspondence during my absence, if you find anything requiring my immediate attention mail it to me at Virginia City. You may use your discretion in answering all other communications.'

"Well, Orrin had been gone about a week when I did use my discretion as I was authorized to do. On entering the office on this morning I found, among the other mail, a big fat document with a half dozen three cent stamps stuck on it. It was such an important looking letter that I came near forwarding it to Orrin without opening it. If I had done so my job would not have ended so abruptly; but I finally broke the seal and read it. It was from a broncho buster and cow puncher over Austin way, giving Orrin a long-drawn-out account of the depredations of a band of rustlers and horse thieves among the cattle men of that locality and demanding that an immediate stop be put to their activities, and that the thieves be arrested and punished. He evidently thought that Orrin represented the Sheriff, District Attorney, Judge and Jury in the premises. Instead of throwing the fool thing in the stove or pigeon-holing it until Orrin's return, I, like a darn fool, sat down and wrote something like this fool answer:

"'Dear Sir: Answering your very important communication regarding the losses that you and the other cattle men in your locality have sustained through the unlawful appropriation of your horses and cattle by the wicked band of dishonest rustlers and thieves infesting your ranges, I am sorry to inform you that the Secretary of State has been suddenly called to Washington on business which concerns the National Government. An emergency has occurred through which the United States is likely to be involved in a war with Europe. It is a situation that needs to be handled delicately and wisely and it will require exceptional statesmanship to handle it. Secretary Clemens has been selected by the President of the United States as the one statesman best qualified to represent the Government in this critical situation and there is not a doubt but that he is now somewhere on the Atlantic Ocean speeding on his way to the Court of St. James. I have wired the State Department at Washington to transmit to him by cable all the facts contained in your letter. When he receives it you may rest assured that he will take immediate steps to have the scoundrels who have been robbing you promptly rounded up and hung.'

"That confounded letter was published in the *Austin Sage Hen,* and when Orrin read it he was so mad that he never even waited to see me, but wrote a note, saying:

" 'Young man, you had better go back to Missouri, for if I find you in my office when I get there I am going to thrash you to within an inch of your life.'

"This is one reason, Bill, why I hesitate to accept your offer. Another reason is that it will break up my work as a newspaper man. This is work that suits me and for which I think I am adapted, and upon the whole I think I had better stay on the job here with Joe."

"Sam, your duties as my secretary will not in any way interfere with your work as a writer. On the contrary it will open a broader field for the exercise of your literary talent than you have here as a reporter on a newspaper. As for taking the matter under advisement, that is altogether unnecessary and I hope that you will accept the position at once. If before I leave for Washington, which will not be before November, you find, for any reason, that you are not satisfied, you can resign and everything will go along as it is going now. In the meantime your work on the *Enterprise* will not be interrupted in any appreciable degree. With the exception of an hour and a half a day in assisting me with my correspondence your time will be your own."

"Very well, Senator, with this understanding, I accept your offer. When shall I report for duty?"

"Oh, any time to-morrow after eight o'clock a. m. that suits your convenience."

"All right then, I will be at your office just after lunch."

The Senator then rose from his chair and, thanking Sam, shook hands with him and left the office. When it became known that Sam had been appointed Stewart's secretary, his friends, wherever he met them, would raise their hats, bow and greet him as Mr. Secretary instead of in the old friendly way of "Hello, Sam, how are you?" Then chuckling to themselves, they would pass on, as though they had played a good joke on him.

At first Sam paid no attention to this change in the demeanor of his friends, returning smile for smile and bow for bow when he met them and continuing on his way to Stewart's office, where for the next month he discharged the duties of secretary to the entire satisfaction of the Senator. But when, at the expiration of that time, the boys in the office got into the game and began quizzing him, as only a lot of mischievous fun-loving type drivers could, it got on his nerves and he began to lose interest in his job.

Salty Boardman started the ball rolling one night when he entered Sam's room for copy. Spying Sam's hat on the table, he picked it up and after examining it, placed it on his head and, grinning from ear to ear, said:

"By golly, Sam, this is a mighty nice hat, but I guess you won't want it much longer, and as it is just my fit, I wish you would give it to me when you are through with it."

"Very well, Salty, when I discard that hat, you will be welcome to it, but why do you think I won't want it much longer?"

"Why, I thought as your head was growing so fast you would soon have to get a hat big enough to fit it."

Then Salty bolted out of the room just in time to save himself from being hit by a paper weight which Sam hurled after him.

"Say, Sam, what are your duties as Senator Stewart's private secretary? Is there anything secret about them?" another of the boys would ask.

"Oh, no, nothing at all. I am simply helping him with his correspondence here in order that he may depart for Washington with a clean slate."

"Oh, is that all? I had the impression from what the janitor at the court house told me, that you were just an all round office boy; sweeping up, running errands and things like that."

This continued chaffing finally became so annoying that Sam began remaining away from the Senator's office for days at a time. These frequent failures to report caused Stewart to lose patience and he began to complain, not only to Sam, but to their mutual friends, regarding his dereliction of duty. When he was told of this outside faultfinding, he at once determined to quit, and writing his resignation, he proceeded with it to the office. When he entered Senator Stewart looked up and without speaking motioned him to a chair, then resumed his writing. After regarding him a few minutes Sam hailed him with,

"Not feeling well this morning, Bill?"

"If my feeling well or otherwise has any interest for you, no," snapped Stewart, "I am not."

"I am sorry, Bill. What seems to be the trouble? Anything gone wrong?"

"Your conscience should answer that question, young man. Sam, as a reporter and an all round newspaper man you have no peer, but as private secretary to a United States Senator you are a complete failure. I think, therefore, that the relations existing between us had best come to an end."

"Bill," drawled Sam, "I am greatly pleased that you have been the first to broach this subject. I discovered some time ago that the situation was an impossible one for me, for I found that although you were a howling success as a lawyer here in Virginia City you wasn't worth a damn as a United States Senator, so I decided to hand in my resignation. Here it is, Bill."

And Sam, after making the Senator a sweeping bow, slammed the door and departed.

Dan and I Witness the Piute Card Game

ONE night as Dan McQuill and I sat talking, after our copy for the next morning's issue was all in, he said to me:

"What do you say to taking a walk with me to-morrow afternoon?"

"Well, I don't care, Dan. Where do you propose going?"

"Did you ever see the Piute's game at cards played?"

"No, I never did. Is there anything peculiar about the game?"

"Yes, it is peculiar in every way. So peculiar that no white man has ever learned to play it. I, as you know, speak the Piute language and they are all my friends. They are always ready and willing to render me a service or to give me information regarding themselves or anything except their game. I have tried time and again to get one of them to explain it to me, but they have always been met with 'This Injun game. Tell white man Injun go broke.'

"If you will go with me to their camp on Cedar Hill to-morrow you will see the game for yourself. I saw two Digger Bucks from California on their way to the camp this morning and they will stay until they are skinned down to their breech-clouts."

"All right, Dan, we will go and watch the *modus operandi* of the skinning."

The next day when we started on our walk Dan went into the store of Fusier & Company and bought a small package of dried shrimps and put them in his pocket.

When we reached camp we found a dozen or so Indians including the "Walla" bucks, surrounding a big blanket, which was spread upon the ground, and old Piute John was sitting on a soap box shuffling a half dozen decks of cards together, preparatory to dealing them. When this process was completed, he scattered them over the blanket. He then took a stick, with a crook on one end, in his hand, straightened up and grunted "Hiskee."

This was evidently the signal for beginning the game and the players would place their bets by simply tossing their money among the cards, without a seeming choice of any particular one of them. When the bets were all in, John began moving his crook slowly around above the cards, at the same time, seemingly, came from the pit of his stomach, "Sargwo, tiley, jovey, nunkey."

He then began drawing in the bets from all over the blanket, paying one here and there until no money remained on the blanket. Then without in any way disturbing the cards, uttered the word "Nippa," rose from his box and came to where Dan and I were standing and extending his hand, greeted Dan with:

"How, Dan, who other white man?"

Dan answered him in Piute, telling him that I was his friend and partner. John then shook hands with me and, pointing to his tepee, said:

"Make heap big eat, deer meat. Come."

Dan told him that we were not hungry and, handing him the shrimps, said, "Heap good chuck, John. You take um."

John took them, but when he opened the bag and saw its contents, with a horrified, "Ugh!" he threw it from him and then, with an angry frown, turned on Dan with:

"Whas a matter give Injun scorpium bug. Glass hop belly good, clicket bug, me like um." Then wrapping his blanket around him he angrily strode away.

Just as John finished dealing the next hand and had pronounced the word "Hiskee," Billy Brimacomb and Abe Brill, two members of Virginia City's sporting fraternity, joined the onlookers.

"Hello, Dan," ejaculated Billy. "You fellows trying to bust the bank?"

"No," replied Dan, "busting banks is not along the line of our business. Maybe you and Abe had better try."

"Well, here goes," said Billy, tossing a half-dollar on the blanket. The coin fell on its edge and, after spinning round a moment, stopped on the ace of diamonds. After going through the usual motions and grunting "Nunkey," John began drawing in his winnings and paying his losses, winding up by adding four half-dollars to Billy's one.

"Well, I'll be darned, boys, if I haven't called the turn. I'll just let the money stay where it is."

"No takum money?" asked John. Billy simply shook his head. Old John once again said "Hiskee," and the game went on. At the end of this hand Billy's luck deserted him. John not only raked in his bet, but demanded another dollar from him. Billy threw the dollar over to John and turning to us, laughingly said:

"Boys, this beats a hogging game of faro all to hell. Let's git."

By this time the two "Wallas" and most of the other Indians had transferred their possessions to old John, and as we had witnessed what we came to see, Dan and I went back to town with the two sports.

"Uncle" Jimmy Fair's Peculiarities

In the summer of 1872 Con. Virginia stock was listed at from twelve to fifteen dollars per share. In less than one year it jumped to five hundred dollars, turning a score of men of very moderate fortune into millionaires. The two most prominent of these were John W. Mackey and James G. Fair.

These two were associated in the management of the mine. John Mackey was held in the highest esteem by all classes along the Comstock, and deservedly so, for the sudden acquisition of great wealth did not change him from the plain, quiet, unassuming gentleman. He had a great human heart in his breast. If a man in his employ became crippled, or otherwise incapacitated for work, John would always find an easy job for him in the mill or around the works.

Mr. Fair was the antithesis of Mackey. He had no use for crippled men, saying that he did not want any but whole men in the mine or around the works.

He knew every member of his working force and gained great prestige by the fatherly interest manifested in them and their affairs. They were all his "sons" and he was their "Uncle Jimmy." Uncle Jimmy's interest in his men was not by any means a feigned one, but sympathy and good fellowship were altogether lacking. He had no confidence in the men's honesty, but suspected them all of "high grading" or shirking. On one occasion, while he was sitting in the office of the secretary, one of the miners entered and approached the desk to receive his pay check. When the secretary handed him the check he turned to leave the room, but before he reached the door, Mr. Fair hailed him with:

"Hello, Tom, what's the rush, my son? You needn't be in such an all fired hurry to cash that check. The Bank of California is still solvent and you have all the time you want between now and tomorrow morning to feed your money to the 'Tiger.'"

"If the 'Tiger' thinks he's going to get any of this money, Uncle Jimmy, the striped rascal will be fooled a bunch. I have a wife and two children to care for and every dollar I earn goes into my home."

"Do you own your home, my son?"

"Oh, yes, I do now, Uncle Jimmy, but on account of our living expenses, doctor bills, taxes, interest, clothing, the furniture and my wife's piano, it took me a long time to pay for it out of my earnings, but, now, thank God, I have a title to it free from incumbrance and if I can keep my health and hold my job, I hope to be able to put by something for the future."

"How long have you been at work in the mine, my son?"

"I have held my job in the Con. Virginia for five years."

"And in those five years you have supported yourself and family, bought, furnished and paid for your home with your wages as a miner in the old Con. Virginia? That is certainly a fine record, my son, and I tell you right now that you won't have to lose any sleep about holding your job. I only wish we had more men like you in the mine. Well, I won't detain you any longer, Tom. Remember me to your wife and babies and don't fail to come and see me occasionally."

After Tom left the office, Mr. Fair turned to the secretary and said, "I wonder where that fellow has his reduction works located."

"Why, Col. Fair, do you think that Tom is a high grader? I have known him ever since he began work in the mine and have always regarded him as a straightforward, trustworthy man and I still have that opinion of him. I am confident that you have no reason for mistrusting him."

"Well, maybe I haven't, but he is a Cousin Jack, [i.e., a Cornishman] and I tell you there is not a damn one of them that won't bear watching when it comes to high grading," sneered the colonel.

On another occasion, while inspecting the works, Uncle Jimmy approached one of the men sorting ore and cordially greeted him with, "Good morning, my son, how is the ore looking this morning?"

"The ore in the pile would be hard to beat, Uncle Jimmy, and there is more of it than usual. Some of it shows a lot of free gold and a man could make big wages pounding it in a mortar."

"Well, I must be getting along. So long, my son. Don't work too hard."

The colonel then went to another man, and after talking with him a while asked, "What do you think this ore will pay in the mill, my son?"

"I am not competent to answer that question, Mr. Fair. When the boss put me on the job he told me to save a certain kind of rock and to throw another kind over the dump. I know what I am to save and what to toss away—that is all."

After questioning a number of the other sorters,

Uncle Jimmy walked over to a foreman and said:

"Tom, do you see that red-headed chap over there? Fire him. He knows too much about ore to suit me."

There was a rule prohibiting the miners smoking while on shift. Infringement of this rule meant the loss of the job. One evening Mr. Fair entered the engine room and, stepping upon the cage, requested the engineer to lower him to the nine-hundred-foot level. Stepping from the cage at that depth he proceeded to the face of one of the drifts. He greeted the miner working there with, "Slow down, my son, you'll hurt yourself by slinging your pick in that nasty way. Take a blow and let me spell you awhile. Now you just watch your Uncle Jimmy drive this drift ahead."

And taking the pick, the colonel went at the face of that drift with a will and soon proved himself to be a past master in handling a pick. When he laid the pick aside at the end of half an hour, he faced the miner and fairly howled:

"Whooee! That's the dope, my son, to limber up stiffened joints and take the kinks out of a fellow's muscles. My, but I do feel good after that stunt, and I will just top it off with a smoke and go on top feeling younger than I have for five years, so hand out your pipe, my son, and make me happy."

"The men are not allowed to smoke underground," said the miner, "and if one of them was caught at it, it would cost him his job, so I guess you will have to wait until you get on top."

"What has that to do with it? There is no question of you being caught, I am the one who is going to do the smoking, not you."

"That's all very true, Uncle Jimmy, but it's my pipe and tobacco. It amounts to the same thing, don't it?"

"There is not a word in that order about your pipe and tobacco. It says that you shall not smoke in the mine and that is all so. Don't keep me waiting, but pass along the dudeen."

The miner, being at last convinced that he was not risking the loss of his job, handed over his pipe and his tobacco, which the colonel took with a satisfied smile. Filling it with Bull Durham, he sat down on a candle box and contentedly smoked for perhaps another half hour, the while entertaining the miner with anecdotes of the time he spent in mining in Tuolumne County and other localities in California. After finishing his smoke, he arose from his seat and handing the pipe to the other, said:

"I have keenly enjoyed every moment spent with you, my son, and your Uncle Jimmy is coming again, and don't you forget it, but I must be going now. Don't work too hard, and say, if you feel like taking a smoke, take it."

With these friendly words he stepped upon the cage and was hoisted to the surface. Arriving there he said to the foreman:

"John, you know that sawed-off Cousin Jack working in No. 2 gallery at the nine hundred. Fire him! He will set the mine on fire with his damn pipe."

Meeting this same man on the street a few days later, he accosted him with, "Hello, my son. What are you doing here this time o' day? Aren't you working?"

"No, Uncle Jimmy, the foreman fired me."

"Fired you. Do you mean to tell me that John Ricks fired you? What for, I should like to know?"

"I don't know what he fired me for. All I know is that when I reached the shaft the day after you were in the mine, he gave me my time and told me to go to the office."

"Oh, he told you to go to the office without assigning a reason for it, did he? Well, maybe Mr. John Ricks will be told to go to the office himself.

I'll let him know that he can't discharge one of my favorite men without cause. I will look into this matter right away, my son, and see that justice is done you."

He then went on his way, feeling that in this case, the ends of justice had already been attained.

Like Bret Harte's "Heathen Chinee," for ways that were dark and tricks that were vain, this millionaire was peculiar.

Mark Twain and the Goat

WHEN war was declared with Mexico in 1846, Dick Stoker, Mark Twain's character of "Dick Baker" in "Roughing It," was a prosperous business man in a little town in southern Illinois. The call of his counrty was to him like the voice of God, and he sold out his business, enlisted in the army for the war, left home, friends and kindred and his promised bride, and followed the flag into Mexico. He fought in all the principal battles during the war, under both Taylor and Scott, was twice wounded, once at Chapultepec and again at Molino del Rey. He was with Scott when he raised the American flag over the palace of the Montezuma in the City of Mexico. As his term of enlistment had expired, instead of returning to the East, he came to California overland with two of his fellow soldiers and townsmen. He stopped a while in Monterey, buying two burros. One he packed with his tent and household affairs and one with provisions. At San Francisco he crossed the bay with his two friends to the east side and followed on up the trail to Sutter's Mill. On arriving there he found the gold-producing ground located for miles around. He went over to Placerville, then known as Hangtown. Finding the conditions there the same as at Sutter's Mill, he turned back south and west to where Latrobe is to-day and prospected on through Amador and Calaveras counties and finally, on the fifteenth of November, 1849, arrived on Jackass Hill. There were about fifty people prospecting for gold. Stoker at once went to work and struck gold in Indian Gulch, along in December, 1849.

Among the things he brought from Mexico was the sword he carried all through the war and a big ten-inch horse pistol that he had taken from a Mexican. This pistol carried a bullet as big as his thumb, and when discharged sounded like a six-pound cannon.

In February, 1850, he built a cabin. In this he partitioned off a little place and covered it with some kind of cloth in the corner. This he called his wardrobe. There the pistol and the sword hung until Mark Twain arrived on the Hill. Mark Twain arrived at the cabin on the twenty-sixth of October, 1864. About a month later he discovered the pistol and sword hanging in Dick's wardrobe. He and I were alone in the cabin at the time and he turned to me and said, "Is this the kind of tools you gold diggers use in digging out the precious metal here?"

I said to him, "No, Sam, they belong to Dick Stoker. The sword was the one that he carried all through the war with Mexico."

Said he, "Do any of you people understand fencing or swordsmanship?"

Said I, "No. Only Stoker. What he knows about it I don't know."

He said, "We will organize a company of swordsmen on the Hill and we will call them the Jackass Squad of Invisible Swordsmen. If any of you fellows wish to learn swordsmanship, I will see what I can do in teaching you."

So the squad was organized and every morning after that he would take us out and practice us at swordsmanship, he alone having the sword. The others were armed with hoops and sticks, anything they could get to fence with. He always would end by disarming the whole squad and then he would give us a tongue-lashing. Stoker was the only one that ever stood against him for any length of time.

Mr. Carrington, a man from Hannibal, Missouri, with his family, lived on Jackass Hill. They were in strong sympathy with the southern people then at war with the north. George Carrington had a great big Australian billy goat that he had trained to harness.

One morning after Sam had disarmed the squad, the old billy goat was standing looking on and Mark Twain noticed the goat and went over and said: "General, you have built up quite a reputation during this war. Now I think there is no question but that you know how to plan a battle and to carry it to a successful issue, but as a swordsman, I think you have been very greatly overrated. Now I'm going to try you out." And he went up and presented a sword and said: "On guard." And made a pass at Beauregard. Beauregard, the old general, was ready for the fray and reared up on his hind legs and hit the old sword with his horns and knocked it about twenty feet. Then he gave Mark the other horn on the side of the head and knocked him flat. Then Mark got on all fours following the old sword. It was the only time I ever heard Sam swear in earnest. He got to the old sword, brought it into the house, threw it on the floor and went out where Mr. Carrington had been watching the battle and said to him, "Carrington, what will you take for that confounded old goat?"

Said Carrington, "What do you want with him, Sam? You want to practice fencing with him?"

"Practice fencing! No, I want to kill him. He is a menace to the limbs and life of the people on the Hill. You know there are a good many children here and he is apt to kill or maim some of them very badly."

"Why, Sam?" said Carrington. "Don't you know that self-defense is the first law of nature and that you started the trouble with Beauregard? He was only working under his own right."

"The idea," said Sam, "of self-defense of a confounded goat against a man."

"Well, all the more reason," said Carrington. "He had a perfect right to defend himself and that is all he did."

Sam then came into the house and I had a bottle of Triple Horse Medicine and I made him lie down on the bed and I rubbed his neck for one-half hour, but he never bothered old Tim again. On the contrary he got friendly with General Beauregard and he and George trained another goat to work with Beauregard in harness and they used to have good times out gathering pine cones, sticks, pine-nuts, and kindling for the mother and the cabin.

Mark Twain at a Wedding

ONE Saturday morning when I entered the cabin I found Sam shaving himself, while all his good clothes were spread out on the bed.

"Hello," said I, "what does this mean, going somewhere?"

"Yes," he answered. "I am going to Sonora and will go to church to-morrow with brother Masons. Now if you want to see me dressed up and all that, you had better go along. You have nothing to keep you here so come along and go with me. It is bright moonlight and we can come home after supper."

"All right, Sam, I'll go with you."

So as soon as we got ready we went over the Hill to Sonora. After looking at the procession we had dinner with the Masonic Fraternity at the Victoria Hotel and I went along as Sam's guest. After dinner we went shopping in nearly all the stores in Sonora and bought necessary articles that Sam wanted. Along towards evening we concluded we would go home and were on our way to the City Hotel to get ready. While we were walking down the street the Reverend Mr. Croche joined us and took an arm of each. He said to us, "I'm glad to meet you gentlemen to-day because I want you as a witness to a wedding."

We went to the hotel with him and we found about a dozen people there. Some we knew and some we didn't.

Sam says, "It is unusual for people to witness a wedding, isn't it?"

"Oh," said Mr. Croche, "this is a different thing altogether. The man to be married is a woodchopper from Horse Shoe Bend who has bought a wife for twenty dollars and they are to be married in the City Hotel Parlor."

After the ceremony we kissed the bride all around and stood there congratulating them for perhaps a half hour. Then Sam and I adjourned to the hotel office to get ready to go home. The bridegroom was brought down to the barroom and was there until nearly midnight and pretty drunk. He went upstairs and wasn't gone but a few minutes before he came tearing down like a whole band of mustangs, slapping the table with his hat and hitting things with his fist and going around making a great fuss about what he found.

Sam said to him, "Tom, what's the matter with you?"

Said he, "I'm mad, that's what's the matter with me and I'm spoiling for a fight."

"Well," Sam says, "what are you mad about?"

Said he, "I went up into the parlor and I found another fellow hugging my wife."

"What did you do about it?"

"Well," said he, "the way I slammed that door, they knew mighty well that I didn't like it."

Sam says, "Did you know the fellow?"

"Yes, I knew him. He is her former husband and he followed her from San Francisco."

The woodchopper didn't go back any more and the next morning the former husband took the lady to San Francisco.

The woodchopper, Tom Haslam, went back to Horse Shoe Bend with his ax on his shoulder to chop wood again and let his wife go.

Truthful James Spins a Yarn

BRET HARTE ON HIS UPPERS

ON a dark, stormy night in December, 1855, as Jim Gillis sat by the fire reading in his cabin, with the wind blowing a hurricane and the rain coming down in torrents, he suddenly heard the tapping of some one loudly rapping, rapping on his cabin door. At first, like Poe, he thought it was the wind: "Only this and nothing more," trying to blow the hinges off his door. Then there came a mighty whack, causing the door to rattle and crack.

"What in thunder are you doing out there? Trying to knock my house down? Come in, you chump, the door ain't locked," yelled Jim.

At this somewhat rough invitation a thinly clad young fellow, soaked to the skin and shivering with cold, entered the room and asked for shelter.

"Now, who the devil are you and what are you doing out at such an hour in a storm like this? Sheriff chasing you?"

"N-n-no. I a-a-am —"

"Shut up," snapped Jim. "Get over by the fire there and take off your wet clothes. I'll get you some dry ones." Jim then went into his bedroom and returned with a suit of dry underwear and a coat. "There," said he, "dry yourself with this towel, then get into these and you will soon be more comfortable."

After the young man had put on the dry garments he said, "I don't know your name, sir, but I assure you that I am deeply grateful for your generous kindness to a miserable half-drowned stranger. I would surely have perished in this awful storm had you refused me shelter."

Here Jim interrupted him with, "Oh, let up on that grateful stuff. I would not turn a mangy dog out on a night like this. Sit up to the table and put some grub into your stomach, then you may, if you feel like it, give an account of yourself; who you are, where you are going and why. My name is James Gillis."

After satisfying his hunger, the young man told Jim that his name was Francis Bret Harte and that he had aspired to make himself a name as a writer and thought the mountains of California, and their grand scenery and many legends of the older days would afford him inspiration for his work. "I have been," said he, "at Westpoint, Calaveras County, trying to attain the goal of my ambition, but up to the present time I have met with failure and disappointment. I have disposed of a few stories and one or two poems in the East, but the most of them have been returned marked 'Not available.'"

He then told Jim he was on his way to San Francisco, where he hoped to secure employment on some one of the newspapers of that city, whereby he would be able to earn enough to tide over the time it would take to revise and rewrite his rejected manuscripts.

"And you know, Mr. Gillis, I shall find in the libraries of the city the works of all the great authors of both poetry and prose and by carefully reading and studying them I hope to get a real inspiration for my work which will help me to 'win success from failure' and make for me a name."

"Well, Mr. Harte," said Jim, "I hope that your ambition may be gratified to the fullest extent, but come, let's get to bed."

The next morning just as the stage came into sight, Jim said, "Bret, it's going to be a cold ride to Sonora this morning, so I will lend you my overcoat. You can leave it with the clerk at the stage office."

"No, Mr. Gillis, I have no money to waste on stage rides, so I will just have to foot it."

"What!" cried Jim. "Foot it over to Sonora in all this wind and rain and slush! How much money have you got, anyhow?"

Bret put his hand into his pocket and pulled out four half dollars, two quarters and a three-cent piece.

"Why, you poor young devil," said Jim, "is that all you've got? Do you expect to get to San Francisco on two dollars and fifty-three cents? Here, add this to that pile you've got in your hand," handing him a twenty-dollar gold piece, "and I think you will pull through all right. Now good-by and good luck to you."

HARTE AND TRUTHFUL JAMES

Bret's first newspaper work in San Francisco was on the *Times and Transcript* as a reporter, but he subsequently got a "case" on the *Morning Call* where he came in touch with Jim Townsend, who was known by nearly all the printers on the coast and had the reputation among them of being the champion of all liars. Lie is defined by Webster as "uttering a falsehood with the intention to deceive." According to this definition, Jim was not a liar of any sort whatever. He was just a story-teller with a wonderful imagination and constructive powers. His yarns were spun with no intention of deceiving, but for the amusement of his hearers. He was one of the best story-tellers I ever listened to. Had his adventures on land and sea, his miraculous escapes from impossible situations, such as being caught in the tentacles of an enormous devil-fish in the Indian Ocean and saving his life by cutting himself free with a small pocket knife, and many other hair-raising tales of escape from disaster and death, been published, he would to-day, instead of being dubbed the greatest liar in the world, be considered one of the great humorists who have filled the world with smiles. This man, J. W. E. Townsend, was the original of Bret Harte's "Truthful James," not my brother, James N. Gillis.

TRUTHFUL JAMES SPEAKS

After "Copy" was all in type and the forms locked, the printers in the *Call* office frequently gathered around the stove in the composing room to while away an hour or so telling stories, discussing the topics of the day, etc.

Sometimes during a lull in the conversation, one of the boys would exclaim, "Come, Jim, spin us a yarn." He would then relate some gruesome experience of his own that would make his audience grow pale and cause them to silently stare at Jim in horrified wonder, or tell them of some impossible predicament in which he had been caught and from which he had extricated himself in an impossible way, which was so intolerably funny that the boys would jump from their chairs and rush from the room almost strangling with laughter.

One of these stories was of being cast upon an uncharted island whose inhabitants were all women.

"Every one of them," said Jim, "wanted me for her husband and they nearly pulled me in pieces, trying to get possession. A dozen of them would grab me at once and pull and tug and haul, jerking me this way and that, until every joint of my body seemed to be dislocated, my bones all broken and my muscles stretched so tight that they would have made good fiddle strings. I had almost given up all hope of coming out of the scrimmage alive, when they began a battle royal among themselves.

"Screaming like panthers, they went at each other with finger nails and teeth, clawing, scratching and biting like a pack of hungry wolves.

"Well, boys, the upshot of the whole business was that, to make peace and save my own life, I married the whole bunch. Say, you may talk about two old Jewish kings, David and Solomon, why so far as wives are concerned I had them both beat to a frazzle. They had only two hundred apiece, while I had a thousand.

"Well, I finally escaped and got back to civilization, but I am willing to bet there are more people on that island named Townsend than in all the world put together."

At this point in the story, after the merriment had subsided, Bret Harte, who had been silently listening, said, "You have not told us how long you were on the island, Jim, nor how you escaped therefrom. Go on and finish the story."

"I thought I'd leave that part out, Bret, fearing that you all would think I was lying, but as you have asked me to, I will tell you. I was on the island a little more than two years. During that time I was loved and coddled and petted so much; changed my residence so often; was stuffed with so many kinds of grub and had to wash it all down with a damn, vile homemade mixture they called beer, that I became afflicted with stomach trouble, which made all my food go back on me and I lost flesh and weight so rapidly in my last six months' stay on the island that I became so thin the sun would shine right through me, and so light that I had to tie a big rock around my waist to hold me down. But that stomach trouble which I thought at the time a great affliction was a 'blessing in disguise,' for it was the means of my escape from, not the island alone, but from my wives as well.

"One day I was sitting in the shade of a big palm leaf I had gathered braiding a small rope of sea grass when I saw two of my wives approaching with a devilish concoction they called siwash, meaning a tonic. I never took it without a struggle, but they got it down my gullet all the same. They would jump me, throw me on my back and while one sat on my chest and held my nose, the other would stick the small end of the gourd as far down my throat as she could and keep tapping the big end with a small stick until the gourd was empty and the stinking stuff safely coiled in my stomach where it lay, like a wet gunny sack a few moments and then begin to unwind and try to crawl out again.

"On this day, before beginning my braiding, feeling that I would relish the milk of a cocoanut, I tied a rock to my waist just heavy enough to overcome the attraction of gravitation and jumped to the top of a big cocoa tree and gathered a couple of the nuts. When I saw my wives approaching with my midday meal of siwash, panic took possession of me and grabbing my big palm leaf, my rope and one of my cocoanuts, I jumped for the top of a tree growing near the beach, intending to remain in hiding among its branches until nightfall, hoping thereby to escape that hellish brew forced down my throat. But owing to the small rock around my waist I miscalculated my jump and was carried fully twenty-five feet above the top of the tree and went sailing before the wind out to sea. When I realized my position panic again seized me and I had about given up all hope of escaping alive when, looking back to the island, I saw my two wives on the beach wildly gesticulating with their hands and arms, beckoning me to come back and when I saw one of them stoop and pick up the gourd, containing the devil mad mixture from the beach and begin frantically waving it back and forth, I at once became reconciled to my situation and was made almost happy at the thought of drowning.

"I reached the water in a horizontal position. After lying there a few moments bobbing up and down like a cork I began drifting before the wind, going round and round like a toy boat without a rudder. I then began trying to remedy this roundabout motion so that I could make a straight course ahead. After a few trials I found that by steering with one of my feet I could drift before the wind bow on. My success caused a hope to spring up within me that with my big palm leaf as a sail, I might reach a place of safety somewhere.

"I at once began to carry out my plan by adjusting my rock as ballast to keep me on an even keel and by the time my preparations were completed, the wind had increased to a very stiff breeze. So, hoisting my palm leaf I sailed away at a good fifteen knot clip due east, for some unknown port.

"Just as darkness was falling on that first day of my voyage I became very drowsy, and after securing my sail so that it would not go adrift, I turned over on my back and went to sleep. I slept through the whole night like a little baby. I woke next morning, just as the sun was rising, greatly refreshed and feeling like a new man.

"Noting that my course had not changed during the night, I took a few swallows of milk from my cocoanut, hoisted my big palm leaf and sailed away eastward. During the night the wind had slackened to about a five knot breeze. I had been sailing along at this rate for something like three hours when it died out almost entirely so that I had barely steerage way.

"A little later the wind left me altogether and there I was, insofar as I knew, in the middle of the

Pacific Ocean with nothing to subsist upon but one cocoanut; in a dead calm which I knew might last for weeks, and my heart sank within me as I contemplated the gloomy prospect ahead of me.

"As I lay there bemoaning my unhappy fate and conjuring up all sorts of plans to escape I began to wish myself back with my wives on the island, but, when a vision of that damn gourdful of siwash, waving back and forth, in the hands of my wife, appeared before me, a deadly sickness overcame me and I began retching and gagging until everything within me, as well as the siwash, went into the sea never to reappear. While I was still in the throes of my death agony, as I then thought, I felt a sudden jerk on the line around my waist.

"Immediately thereafter the snout of a big swordfish shot out of the water right at my side. The son-of-a-gun had swallowed my rock and was now trying to dislodge it by opening his jaws and snapping them together again, at the same time jerking his head first to one side, then to the other.

"As I had on several occasions saved my own as well as the lives of my shipmates by acting promptly and quickly in an emergency, I was in a measure, prepared for this one. So, gathering the slack of my rope, I took two half hitches around the snout of the fish and pulled them taut, thus eliminating the danger of being caught between his jaws when he snapped them together. When, after accomplishing this feat so that I could prevent the fish from diving by jerking the cord, the hope of extricating myself from my perilous situation and saving my life by riding him to some haven of safety came to me.

"I straddled his back and tried to make him start ahead by slapping the side of his jaw with my rope and digging my heels into his ribs. He started all right, but instead of taking a straight course ahead he started in by hunching his back and making a vicious lunge to port and began swimming round in a circle, all the while lashing the water with his great tail until the ocean, for miles around, was covered with foam. I tried to stop him by pulling on my cord but the harder I pulled the faster he went round. I at last got seasick and was about to let go my rope and slide into the sea, when suddenly he stopped and tried to reverse and waltz around the other way. He was balked in this maneuver, however, when the slack was taken up and the cord around his snout became taut.

"When I realized that I had him tied with but a single rope and that I had been pulling one way on it all the time, I felt like kicking myself for being a damn fool in trying to control him with only one bridle line. So, while he was getting his wind for another bucking stunt, I quickly made my braided rope fast to the other side of his snout, thus providing myself with a pair of bridle lines. I then again tried to start him ahead, but finding that he could turn neither to the right nor left, he got sulky and in spite of my jerking and kicking, wouldn't budge an inch but just laid there pawing the water with his side fin and once in a while suddenly bowing his back trying to throw me off.

"At last, becoming tired and disgusted with the way he was acting, I gave him two sharp jabs with the blade of my knife. The effect was almost instantaneous. With a jump that I thought had dislocated my neck, he started racing through the water at a speed of not less than a hundred knots an hour. He was going so fast that if I had not had a death grip on my lines, the head wind created by his tremendous speed would surely have blown me off his back, but by freezing to my lines and bending low over his neck I succeeded in retaining my seat. By sawing with all my might on the lines I finally succeeded in bringing my steed down to, what on land would be a fast trot. Then, of course, the wind died away also and my ride on the back of the swordfish was the most delightful one that could be possibly conjured up by the imagination.

"Smoothly gliding, without a jolt or jar, over the undulating waves of the ocean, mounted on a steed that was all a steed should be, I felt my suffering and hardships were nearly ended and that I should soon be at home with my relatives and friends.

"We had sailed along in this way for several hours and had covered something like two hundred miles when, just before sunset, I spied land dead ahead. I was so filled with joy that I raised my voice and shouted 'Land Ho' with all the power of my lungs.

"When I shouted out 'Land Ho,' that darned swordfish must have gotten wise to the situation at once for cocking his head to one side, he gave it a shake or two and started at full speed for shore. I tried my best to stop him by sawing on my bridle lines, but it was of no use. He had determined to reach that land in record breaking time and 'taking the bit in his teeth,' he tore through the water at a speed of a thousand miles an hour. When we reached the beach his speed was not diminished the fraction of an inch; he just sailed along over the ground until he ran his sword through a big breadfruit tree growing about two hundred yards from the water. Then he stopped, but I didn't until I reached the end of the rope around my waist, the other end of which, being tied to the rock in the fish's belly, so when I was catapulted from his back that darned rope jerked me shut like a jack-knife when I reached its end and I fell to the ground with the wind all knocked out of me and consciousness gone.

"When I came to myself I was still lying all doubled up on the spot where I had fallen. I got to my feet and started to walk to learn whether I had received any serious injury, but at my first step I sank to my knees in a thick bed of leaves covering the ground. It was night and darkness around me was absolutely impenetrable. It was so dark, in fact, that I began to fear I had gone blind.

"This fear, however, was dispelled a few moments later when I saw the light of thousands of fireflies flashing all around me. Relieved of this fear, and finding I had not been injured in any way, also knowing that I could not accomplish anything until daylight, I decided to remain where I was until morning. So, raking a big hole in the thick bed of dry leaves covering the ground, I lay down in it and drawing the leaves around me, sank into a dreamless sleep and never in this world had man a more luxurious couch than that bed of soft, dry leaves.

"I was awakened the next morning by the chattering and chirping of birds all around me. Opening my eyes I saw them — birds of all the hues of the rainbow, nearly splitting their little throats with song as they flitted from branch to branch among the trees.

"As I lay there with closed eyes listening to their happy warblings my own heart became filled with thanksgiving for my rescue from the sufferings and perils so lately surrounding me. As I was unable to join in their song of praise, I began to voice my gratitude by repeating the Twenty-third Psalm, which I had learned at Sunday School when a boy. 'The Lord is my Shepherd. I shall not want.' I had gotten thus far in my devotions when I was startled out of my prayerful mood by a harsh voice above me rasping out:

"'Who the hell are you?' This rough hail from, as I first thought, a human voice set my heart to thumping like a trip hammer, but before I had time to get really scared it came again with, 'Go wash yourself, you swab.'

"On looking where the hail came from I saw a big green parrot sitting on a limb not ten feet from me with his head cocked to one side, evidently sizing me up. When I noticed the friendly way in which the little creature was acting while sitting there preening himself and looking me over, I felt that he had been sent as a companion and friend and my heart was full of thankfulness as I got to my feet and started for my beautiful visitor with my hand extended, at the same time trying to coax him with, 'Polly, Polly, pretty Polly, come Polly.'

"The parrot manifested his pleasure by climbing to my shoulder and while fondling my face and neck with his beak softly chuckled, 'Poor Jack, Poor Jack.'

"Now that Polly and I had entered into a covenant of comradeship I saw no cause for hurrying my departure. I decided, therefore, to remain for a few days at least where I was. So with Polly on my shoulder I went over to where the swordfish was lying dead and began to look him over. There he lay with his snout protruding fully six feet through the tree, while the ground facing the beach for a distance of at least sixty feet was hollowed out in a half circle into a deep trench by his mighty efforts to free himself. As I stood reproaching myself with ingratitude in not trying to help the poor fellow in his dire extremity my eyes filled with tears and I began to sob. I was aroused from my abstraction when I thought of my rope and as I knew that I should need it later on I decided to secure it before going back to camp.

"I had coiled about two fathoms when I found as the other end was tied to the rock inside the fish, that I would have to cut him open in order to get the remaining twelve feet, else cut my rope in two. Now I hated like the devil in any way to mutilate the fish. I also hated the idea of losing half my rope. I had been debating with myself which course to pursue for some ten minutes when Polly gave me an inspiration by breaking in with, 'Aw hell, cut it out.' Now although I knew that Polly's words bore no reference to either the fish or rope I at once determined to act in accordance with his suggestion and recover my rope in the only way possible. After two hours of hacking and sawing with my pocket knife I succeeded in cutting through the file-like skin and tough flesh of the fish and prying him open with a dead limb found on the beach.

"The first thing I took from the flotsam and jetsam in the monster's maw was a carpenter's crosscut saw, then came a belaying pin, a marlin spike and a broken capstan bar. After these, two binnacle lamps and a ship's compass. I drew from below these articles a crumpled southwester hat, an oil canvas coat and trousers and a pair of rubber sea boots and a magnifying glass.

"As it was growing towards sunset, I concluded

to knock off for the day and get something to eat. So gathering my treasures Polly and I started for camp. After a hearty meal of bananas and cocoanut milk with the parrot perched on a limb just overhead, I burrowed into the leaves and we were soon in the land of dreams.

"I was awakened the next morning as I was the morning before, by the songs of the birds all around me.

"After a hurried breakfast of mangoes and breadfruit I began making preparations for my expedition into the forest planned the day before. I had selected the saw, the rope and the knife as the only equipment needed for my trip and with Polly on my shoulder, was just about to start when Polly screamed into my ear, 'Port your helm. Hard a port.' Glancing over my shoulder there, not more than twenty feet behind me, stood two gigantic naked savages with their spears poised to impale me. Without a second's hesitation I sprang into the branches of the great tree, the spears of the savages just grazing my feet as I rose into the air.

"Then Polly, with frightful screams of anger, flew at the savages like a wildcat, beating their faces with his wings and biting and clawing their bare flesh with beak and talons, until the two, thinking they had run up against the devil, turned and fled into the forest with Polly still clinging to them, biting and tearing their naked skins.

"It was fully ten minutes after the disappearance before the parrot returned to me and I had begun to fear the brave little warrior had lost his life. But at last he lit on the log beside me and dropped an object from his beak which I at first took to be a brown canvas bag but when I took it in my hands I saw that it was a big human ear. The little rascal had actually torn an ear from one of the savages and brought it back to me as a spoil of victory over our foes.

"While I was still gazing at Polly's trophy and wondering at its great size he pulled my sleeve and grated out, 'Bloody pirates. Better beat it.'

"'You are right, Polly,' said I, 'we will beat it and beat it right now.'

"So just at noon of that day, with Polly on my shoulder, I sprang forward into the air and away, away we went, swiftly and smoothly as a bird speeding over the land toward our faraway home in the north.

"Wishing to bring my rate of flying down to a more moderate one, I began experimenting with my palm leaf and in a very short time found, to my delight, that by manipulating it in certain ways I could not only regulate my rate of speed but could also ascend or descend and shape my course to right or left, or in any direction I cared to go.

"Having attained this desirable end, I now reduced my velocity to about twenty miles an hour and gliding along I could distinctly see every object along my route. The beauty of the scenery, the great trees and undergrowth, the infinite variety of lovely tropical flowers, and the gloriously plumaged birds, together with numbers of little monkeys swinging along by their hands and tails from limb to limb among the trees, filled my soul with such indescribable enjoyment that I lost all sense of time or distance in the contemplation of the wonders all around me. I was again brought to a realization of my surroundings by Polly, this time shouting 'Breakers ahead, let go the anchor.'

"Looking to find the cause of Polly's warning I saw that I was heading directly for the face of a great jagged cliff rising hundreds of feet into the air so I at once shortened sail by a quick manipulation of my palm leaf and gently came to anchor within a stone's throw of its base.

"The place where we landed was such an attractive spot that although the sun was more than two hours above the horizon, and I estimated that we had traveled some three hundred miles since breaking camp, I concluded to remain and rest up until the next morning. So, selecting a wide spreading banana tree covered with great bunches of the delicious fruit as our camping place, Polly and I proceeded to pile together dry branches and twigs for a fire and soon had a sufficient number gathered to last us through the night. With this task accomplished as it was still rather early for supper, Polly and I went for a look around to see what we could discover. We first started for the beach which lay sparkling and scintillating in the rays of the setting sun some three hundred yards to the west. Upon our arrival there we found it strewn with thousands of beautiful shells and other countless thousands of crystals and glittering pebbles having all the colors of the rainbow. I gathered a double handful of the finest specimens found among the pebbles and put them in the pocket of my coat, then went on down to the water and went wading and splashing in it like a boy and letting out war whoops of pure joy.

"I was still hugely enjoying the fun when I stubbed my big toe gainst something sharp under the water and reaching down to find what had hurt me I took from the bottom an oyster that was so big I at first thought it was a green sea turtle. When I found that it was an oyster instead I let out a yell that caused Polly to jump fully ten feet into the air, and when he fluttered back to the ground he just sat back on his haunches and fired such a volley of oaths and expletives at me that I began to have my doubts as to whether he was a bird or a fiend. Why, that darned parrot sat there and swore at me in every known language and dialect and in so far as bad words were concerned, had a dictionary beat to a frazzle. When I tried to pacify him by saying in a wheedling tone, 'Pretty Polly,' he repelled my friendly advances with one word, 'Slush,' and waddled off to camp.

"When I got there he was sitting on a limb still pouting. Dropping the oyster to the ground, I lit a match and started a fire, then I proceeded to open the oyster with my sheath knife but found that a bigger job than I had supposed it would be. In fact it proved to be an impossible one, for the shell was so thick and tightly shut that I could not force the point of my knife into the cavity at any point.

"I at last became discouraged and was saying some cuss words myself while looking around for a rock or something with which to smash the darn thing, when Polly dropped to the ground beside me and croaked, 'Saw, saw.'

"'By golly, Polly,' I exclaimed, 'you are a Jewel, sure.'

"'All bosh,' replied the parrot. 'Saw.'

"I complied with Polly's suggestion by sawing through where the shell was hinged together and soon had the two half shells pried apart, and then after broiling two big chunks and sprinkling them with lemon juice Polly and I partook of the most delicious and satisfying supper that I had ever eaten in my life and I have never since had one that tasted half as good.

"After supper, with the parrot contentedly whetting his beak beside me, I lit my pipe and smoked until bedtime. While I was sitting there contentedly smoking, my attention was drawn to a peculiar little bug crawling on my leg and as I had made the study of botany a specialty while at college, I had become deeply interested in all sorts of insects and other small animals. I took him in my hand and began to examine him with my glass. I was closely scrutinizing him and endeavoring to classify him with other saurians I had found along the banks of the Ganges River in Australia when, with a loud click, my glass suddenly lengthened out into a binocle. To say that I was surprised does not convey an idea of my astonishment at this seemingly magic transformation. I was simply lost in wonder but by a careful painstaking examination, I at last found that by pressing on a cunningly constructed spring on the side of the instrument, I could have either a telescope or magnifying glass.

"The next morning as I was gathering wood for fire chancing to glance at the place where Polly and I had opened the oyster the day before, I was surprised to see that every vestige had disappeared during the night, but there in one of the shells lay a magnificent pearl the size of a big walnut. The sight of the pearl sent a thrill of pleasure through me for I knew it was worth a fortune.

"Placing the pearl in my tobacco pouch, I followed a trail leading to the base of the cliff. I learned that my oyster had been devoured by a colony of big red ants who had their nest in a crevice of the cliff. I knew this from the fact that one was posted evidently on guard at each side of the entrance. And of all the ants I ever saw those 'took the cake.' They were simply monsters fully as big as jack-rabbits with mandibles a foot long. And when they started for me with a greedy look in their eyes that plainly told me they intended to put me with the oyster, I turned and fled.

"As Polly and I had partaken of a full sufficiency of oysters for supper the night before, we breakfasted on toasted breadfruit and cocoa milk, topped off with pineapple.

"We resumed our journey and after we had been in the upper air some six hours, I steered for the ground on a slant of about 60 degrees Fahrenheit and a little while later came gently to anchor in a grove of mangoes through which a small stream of crystal clear water was flowing.

"As there was nothing in the shape of food, save mangoes, at the spot where we landed, Polly and I, after quenching our thirst, went further into the forest in search of something more palatable. We had advanced through the dense underbrush only a few hundred feet when we came to an open space which had at one time evidently been a clearing. On the border of this clearing there was a tangle of dark green vines covered with dainty yellow blossoms.

"Stooping to pluck a few of the pretty flowers, without intending to do so I pulled part of the vine from the ground and there on the end of the root I spied a peanut as big as my fist. Dropping to my knees I commenced grubbing at the soft mound under the vines, and soon had nearly a bushel of the big goobers piled on the ground beside me. While engaged in filling the pockets of my coat with the nuts

I chanced to glance to the far side of the clearing when I saw what I thought was a great big oblong bowlder covered with green moss. But on close inspection it proved to be an immense watermelon growing wild there in the woods.

"When I reached the melon I stood for a few moments wondering at its great size. Polly, who was sitting at the upper side pecking at the rind, became impatient and shouted 'Cut it.'

"'All right, Polly,' said I. 'I'll cut it, you bet. But I am going to see how big it is first.' So beginning at the stem end, I took ten long steps before I reached the other.

"Finding the melon to be fully thirty feet long by ten feet in diameter I knew that it would be useless for me to try opening it with my knife. So I took my saw and in about an hour I succeeded in cutting out a plug about four feet square. Then Polly and I got into the hole and filled ourselves with the juicy sweetness of its heart.

"After eating all we could possibly hold, we went back to camp and feasted on roasted peanuts for another hour. As dark was now falling over the land, I decided to camp where we were until morning, so with Polly perched on a limb as usual I rolled up in the leaves and went to sleep.

"Just at sunrise the next morning, after another hearty meal of roasted peanuts, we took to the air again on the last lap of our journey north and at five o'clock that evening, without a mishap, reached San Juan del Sud on the Isthmus of Nicaragua.

"I made my landing just in front of the hotel and going in the office, inquired for the proprietor. 'He stands right before you, sir,' smilingly replied the gentleman at the desk. 'What can I do for you?'

"I then told him my name and related my experiences from the time I was cast upon the Island until my arrival at San Juan, concluding by saying, 'Mr. Mills, as I have no money with which to pay my passage, I should be very glad of a chance to work my way up to San Francisco on some vessel leaving this port and I have come to you, hoping that you will kindly assist me to get that chance.'

" 'You have surely had a rough deal and been up against a mighty hard proposition, Mr. Townsend,' said Mr. Mills, 'and I wonder that you survived to tell of it. Why, there is not one man in a thousand who would have come through such an experience alive. I will be more than glad to help you get passage to San Francisco, but not by working your way. The steamship Pacific is now lying in the harbor waiting for the passengers from the East in transit for San Francisco. Captain Blethen, her master, is a close personal friend of mine and as he is dining with me to-day — I am expecting him momentarily — I will acquaint you with him and when he hears your story I am positive that he will not only afford you passage on his ship but will do everything possible for your comfort and pleasure while on board.'

"Mr. Mills had hardly ceased speaking when a rather portly man dressed in a snug suit of blue, entered the office and in a hearty voice hailed him, 'Hello, Bob. Am I late?'

"'No, Captain, you are just on time. Come over here. I have a friend that I want you to know,' said Mr. Mills.

"He then introduced me to the Captain and said, 'Mr. Townsend has a story that I want you to hear. But come into the dining room; he can tell it while we are at dinner.'

"After I had given the Captain an account of my adventures, and he had put my mind at ease regarding my passage, Mr. Mills ordered wine. The waiter had just filled our glasses when Polly, who had been spooning with a parrakeet in the office, lit on the table, and cocking his eye at the Captain, croaked, 'Ahoy, shipmate. Pass the grog.'

"At this unexpected salutation from the parrot,

the Captain, who had just raised the wine to his lips, gave a kind of choking sob; then snorted into the glass, spraying the wine all over the table. Then, 'Three sheets in the wind, and a bottle of rum,' gurgled Polly.

"'Oh, Lord! Oh, Lord! Hold me, Mills, I'm going to die,' then the Captain sat back in his chair and had such a convulsion of laughter that Mills apparently became alarmed and as there was no water at hand, threw a glass of wine in his face.

"When Polly witnessed this waste of good wine, he wailed out, 'Beat it! Beat it! You're spilling the beer,' then went to crying like a baby.

"When Captain Blethen recovered his breath from the parrot's last sally he rose from the table and said, 'Well, gentlemen, I shall have to leave you and get back to the ship. Mr. Townsend, I should like to have you come aboard as early in the morning as you can make it convenient so that we may have ample time to arrange your quarters before the passengers from the East arrive. They are now two days overdue and as soon as they are on board, day or night, I will hoist anchor and sail.' With these parting words he started for the door, then turned and said with a chuckle, 'Don't fail to bring the parrot along.'

"After giving Polly to Mills and telling him good-by I boarded the gig and was soon aboard the Pacific. Seven days later the ship docked at Washington Street wharf and my journey was ended.

"A few days after landing in San Francisco I got a 'case' on the *Herald* and continued to work there until May, 1856, when the paper was forced to suspend publication because of Mr. Nugent's opposition to the Vigilance Committee."

Here Jim paused and began filling his pipe for a smoke.

"Say, Jim," asked Ward, "how much did you get for your pearl and ruby?"

"Oh, don't ask me, Wardy. it makes me sick whenever I think of them. The pearl was nothing but a piece of gristle that held the oyster shells together. The ants in trying to eat it, after devouring the rest of the oyster, had scraped it with their mandibles and turned it over and over until it was worn round and smooth. While the lapidary was examining the ruby, he happened to get it too close to a gas jet and the darned thing melted. It was just a wad of gum discolored by the action of the air and water there on the beach. I had a cash offer for my jewels while on the trip up from San Juan but declined, as I thought it was far below their value.

"The Emperor of China was a passenger on his way back to Peking from Washington, where he had been on a social visit to the President of the United States. They had hunted big game together in Africa when they were boys. Consequently the relations existing between them resembled those of two brothers. Before leaving China for Washington the Emperor, as he thought, provided an ample sum in cash to defray all expenses and pay for such presents as he desired to make the President and other high officials while on his visit to the capital. Had it been U. S. money it would have not only come up to his expectations, but would have been amply sufficient to buy the whole District of Columbia. There were hundreds of tons of it, all in little copper junkets with a hole in the middle worth about ten cents a bushel. But when he got to Washington, he found himself as good as broke. He couldn't buy anything with them and, because of their bulk, the banks would not give him real money in exchange. Why, it would have taken a wheelbarrow load to buy a ten cent sausage.

"The Emperor, finding himself in this predicament, shipped two hundred and fifty tons of his money back to China by way of Cape Horn, reserving fifty tons for distribution among his subjects in

San Francisco. That was the money he offered in exchange for my pearl and ruby. When I rejected his proposition, he offered to give me twenty-five of his wives to boot. When he mentioned wives, a vision of the island and the gourdful of siwash rose before me and after gagging until my stomach seemed inside out, I told him to take his wives and go to the devil.

"Now, boys, that's the whole story. Don't bother me with any more questions."

When Jim ended his story Dave Hunter staggered to his feet and, slapping him on the back, bawled, "Oh, you darned old liar." Then doubling up, with both hands holding his stomach, he stumbled from the room gasping, "Oh, Lord! Hold me, boys. Hold me, I'm going to die."

The Entomologist and the Yellow-Jackets

ONE night when the Steamer *Chrysopolis* made her landing at Rio Vista on the Sacramento River, an old gentleman with a suit case in his hand walked down the gang plank followed by two deck hands carrying a large iron-bound trunk. Approaching the wharfinger, he smilingly said, "I am sorry to trouble you, sir, but am afraid that I will have to ask some assistance in getting my luggage to the hotel."

"It will be no trouble at all, sir," said that officer; so saying, he directed two roustabouts to carry the gentleman's baggage to the hotel.

After registering at the office he turned to the clerk and asked, "Do you know a young man named James Gillis, whose home is here in Rio Vista?"

"Jim Gillis? Why, yes, everybody knows Jim. His home is on Grand Island, twelve miles up the river, but during this season of the year he spends most of his time here in town, and is liable to show up almost any moment. By George! there he is now. Jim, here is a gentleman inquiring for you."

As Jim approached the desk, the old gentleman, with extended hand, advanced, saying, "Mr. Gillis, I am very, very glad to meet you. My name is Palmer, and I have here a letter for you from Mr. Thomas Wand, of San Francisco, which will doubtless inform you as to my purpose in visiting this section of California."

Jim took the letter, and after reading it, again grasped the hand of the other in his hearty way. "Tom Wand is a very dear friend of mine, Professor Palmer, any one coming from him is most welcome and I sincerely hope that we also may become friends. My time is entirely at your disposal and let me assure you that any assistance I may be able to render you will be rendered willingly and cheerfully. Do not hesitate to call on me whenever I can be helpful in any way."

"Mr. Gillis, I cannot express to you, in words, my appreciation of your kind reception and generous offer of assistance, but I do most heartily thank you. I will take no undue advantage of your offer, nor will I encroach upon your time more than I can help."

"Do not let that trouble you for a moment, sir," answered Jim. "My time is not at present an asset and I am glad of this opportunity to make it one. And now, Professor, I will bid you good night. I will meet you again in the morning."

When meeting at breakfast the next morning, Jim greeted the other with, "Well, Professor Palmer, what have you on the 'tapis' for to-day? Anything special?"

"Nothing of great importance, but I should like, if possible, to secure a small cottage or one or two rooms in a quiet neighborhood where I can enjoy a measure of privacy while at my work."

"I have a friend who will be leaving here in a few days for his old home in Georgia. He has a small, nicely furnished house near the river and I am confident that he will be pleased to have you occupy it during his absence. We will go over and interview him after we have had breakfast."

When Jim knocked at the door of his friend's house, it was opened by a young man of about his own age, whose face was beaming with good nature.

"Why, hello Jim," said he. "I am certainly glad to see you. I am leaving by to-night's boat for San Francisco on my way home, and I had about given up the hope of seeing you. Come in."

"Mr. Dunham," said Jim, "let me make you acquainted with Professor Palmer. Ike, the Professor proposes to remain in Rio Vista for several months. He wishes to rent a house or rooms out of town, where he can pursue his studies without interruption. Knowing that you were about to leave for home, we have called to ask if you are willing to let him occupy your house until you return?"

"Am I willing? Why, Jim, it's the very thing; nothing could possibly suit me better. Professor, as I have completed all my preparations for leaving to-night, you can move in right now if you wish to. As I shall take nothing but my suit case with me, I will give you the key and go back to the hotel with you and wait there for the boat."

After the Professor's belongings were taken to his new quarters the three friends started on a round of visits through the town. Jim to introduce the professor to the people, and Ike to tell his friends good-by.

After seeing Dunham off on the boat, Jim and his friend went to the house by the river and spent the evening together. The Professor detailed more fully his reasons for visiting California. He said that he occupied the "Chair of Entomology" in Princeton University and was engaged in writing a book on a — to him — most interesting subject. As he had read and heard so much of the insect life peculiar to California, he had come to the state for the purpose of collecting specimens and classifying them for his work.

"Mr. Gillis," said he, "the insect family has just as many different traits of character as do the people of the world. Some of them are generous and kind, some are provident, some are wasteful, some gather, some scatter. Some are merciful and some are cruel and vicious. In fact, my young friend, they have nearly all the attributes belonging to man."

"I have always known that certain bugs are beneficial to mankind, while others are the reverse, but I never knew that there were so many dissimilarities among them and I shall be glad to learn more while assisting you in collecting your specimens."

One evening, after an absence of a few days, Jim entered the laboratory and found his co-worker closely scrutinizing some object under a microscope.

"Found something new, Professor?" said he.

"Yes, indeed, James, I have. The fellow I have here is one of the most wonderful beetles that I have ever had the fortune to secure. Take a look at him, my boy, and tell me what *you* think of him."

Jim seated himself at the table and after examining the bug for a full minute, turned, and with a broad grin, said, "Why, Professor Palmer, there are thousands of bugs like this one in my pumpkin patch. It is nothing but a common squash bug."

"Squash bug, squash bug!" almost yelled the Professor, springing from his chair and glaring at Jim. "Why, young man, what is the matter with your eyes? That bug bears no more resemblance to a squash bug than an elephant does to a mouse."

"That may be, but I tell you, it looks mightily like a squash bug to me and I'll be darned if it don't *smell* like one, too."

"Well, we will not discuss the matter further tonight. I think that by morning you may have a clearer vision and be able to see things as I see them."

"I hope that we will not be here in the morning," said Jim. "I want you to go home with me tonight, for I have something to show you up there that will give you the surprise of your life. Yesterday morning while walking over my upper ranch, I found a whole colony of the liveliest insects I ever ran across, and I am positive that while investigating them, you will forget all about this bug."

"This is assuredly good news you bring me, James. Did you observe them closely enough to give me something of a description?"

"Yes, sir, I did. Their bodies are marked with alternate stripes of black and yellow and their wings appear to be of the finest gauze. They have rather large, prominent eyes which seem to be ever on the alert. I spent more than an hour watching them and noticed that they were organized in something like a military way. They live in the ground and when they begin their work in the morning, have one of their number posted at each side of the entrance. These sentries pay no attention to the outgoing members, but all those returning have to undergo an examination at the door before being allowed to pass in. Any one of them returning empty-handed would be denied admittance and evidently ordered to go and bring something home. I will not attempt a further description of them to-night, Professor. You will see them in the morning and will have all the time you care to devote to them; so if you will get the things you wish to take with you we will make a start for home. While you are doing that I will make your skiff fast to the sloop and tow her up to the ranch. To-morrow, after lunch, we will row over the Beaver Slough, where I think we may make some interesting discoveries."

After breakfast the next morning they started out to investigate Jim's find of the day before. The Professor was all aglow with excited anticipations.

When they got to a point about a half-mile

from the house, Jim stopped, and pointing to a small hole a few feet from the path, said, "There's where they live, Professor. Now while you are getting acquainted with the colonists and collecting your specimens, I will go on up to the patch and get a watermelon to take home for lunch."

With these words, Jim started up the river bank at a pretty fast walk. Some fifty yards further along he dodged behind a tree and looking back saw the scientist cut a switch from a willow tree and carefully trim it. After which, he approached the hole and, dropping to his knees, slowly inserted the switch and withdrew it again. Then Jim saw the Professor rise into the air as though he was shot from a catapult, then, "Yeow! Yeow! Yellow-jackets, by G—d!" he howled, and grabbing his hat, he began frantically banging it from side to side of his head in an endeavor to beat off his tormentors, but all to no purpose. The yellow-jackets were not to be denied and kept getting thicker and thicker, until the whole swarm, thousands upon thousands, were over and around him, darting through his guard and sending their spears into bald head, neck and hands. At last the poor old man, with an agonizing yell, took a header into a dense growth of young willows and began crawling through them to the river bank fifty feet beyond, hoping, in this way, to escape.

This maneuver would have been probably successful if the Professor had "laid low" in the willows until the striped warriors quieted down and returned to their quarters, but no, the Professor evidently believed that "He who fights and runs away, will live to fight another day." So he made up his mind to get through the thicket of willows and run from that locality and the yellow-jackets while the running was good, never dreaming that they would be waiting for him on the other side. But, alas! he had barely emerged from the protecting willows when zip! zip! zip! went sting after sting into his bald head. With a yell that woke the echoes up and down the river, Professor Palmer scrambled to his feet and, taking to his heels, sprinted down the path at a speed that would have made him the champion foot racer of the world.

When Jim saw his friend start on his wild flight, with the jackets swarming round him, and heard his cries of pain as they sent their stings into his head and face, his conscience smote him for the cruel joke he had played and he started in pursuit, thinking to overtake the Professor before he reached his boat and get him into the house away from his tormentors.

Before Jim got anywhere near the landing, the old man had got aboard his skiff and shoved off from shore.

"Hold on, Professor," he shouted, "don't go away, the yellow-jackets are all gone. Come ashore and get back to the house." Without paying the least heed to him, the Professor bent to his oars and began pulling down the river as though his very life depended upon getting away without further loss of time.

When Jim found that his calls were unheeded he turned and raced back to his house as fast as his legs would carry him and, providing himself with a bottle of liniment and roll of muslin, hoisted sail on his sloop and followed Professor Palmer to Rio Vista.

When he entered the house he found his old friend lying on a couch groaning with pain. His face was so badly swollen that Jim hardly recognized him as his companion of the morning. His nose and lips were fully twice their natural size, while his eyes looked like two slits in a mask.

Then Jim's conscience again smote him, and he began pleading in a faltering voice for pardon. "Oh, my dear old friend," said he, "please forgive me. I never intended —"

"Young man," interrupted the Professor, "you have been guilty of a wanton, premeditated act of cruelty that would cause an Apache Indian to blush with shame. I had come to regard you, James Gillis, as a young man of high principle, kind heart and generous disposition; I had such confidence in you that I would have trusted you with my life and had come to love you as my own son, but I now find that you are selfish, cruel and vicious, lacking in every essential that stands for a clean manhood. You are no longer welcome here. On the contrary, your presence is offensive to me and I, therefore, request you to leave this house and go the way you came."

"Professor," pleaded Jim, "I know that I am culpable and fully deserve your righteous anger. I have not a word to offer in the way of an explanation or excuse, but I will say that I had no intention of causing you any suffering. I knew, of course, that you would get a big surprise, with, maybe a sting or two, but I had not the slightest idea that the whole army of striped little devils were going to jump you and, when you started racing down the path I thought, of course, that when you reached the house you would go in, shut the door and rid yourself of them."

"What you thought I would do is no excuse

whatever and does not alleviate any of the suffering caused by your cruel trick this morning."

"I am conscious of the truth of your every word, Professor, and I know that you are justly angry with me, but if you will forgive me I promise never again to offend you in like manner. You remember that when His disciple asked Him if he should forgive his brother seven times, Jesus told him, 'Not seven, but seventy times seven times.' Please obey the divine injunction and forgive me."

"Young man," answered Professor Palmer, "that commandment was all very well two thousand years ago, but it won't work in this day and generation. I will forgive you this one time, but no more, and I tell you, that one yellow-jacket nest is all I want, too."

"You will have no occasion to forgive me again, Professor," said Jim. "And now, as we are friends once more, if you will lie down on the couch I will soon relieve you of all pain."

He then saturated the gauze with the lotion and, placing it on his friend's face, left the room, quietly closing the door.

Who Stole the Ham?

UNCLE IKE was a deacon in the African Methodist Church, and to the brothers and sisters of the flock, was like unto a beacon to illumine their way along the pathway of holiness. He could pray longer and shout louder than any other member of the church and his booming Amens at camp meetings and revivals brought many a darky to his knees at the Mourner's Bench. He was a Moses sent to lead his brethren out from the Wilderness of Sin into the Promised Land. But alas! one night as old Satan was roaming around seeking some one to devour, he entered good old Uncle Ike's cabin and led him to a white man's smoke house. He then put a meal sack, containing a big fat ham, on the old darky's shoulder and led him back to his cabin. The next morning, just as Uncle Ike was saying Grace over his "cohn" pone and coffee, the door of the cabin was flung wide open and the owner of the smoke house, followed by his dog, Tige, with a greasy meal sack in his mouth, burst in upon him.

"Why, for de land's sake, Marse Jim," exclaimed Uncle Ike. "Yo' lack skeered me to death bustin' in like dis so airly in de mornin'. Whas de matter, Marse Jim. Sumpin' happen at de big house."

"Something happen! Look at that dog, you black thief."

"Well, I clar to goodness ef dar ain't dat dog Tige, wid a meal sack in his mouf. Whar yo' reckon dat houn' git it, Marse Jim?"

"You know mighty well where he got it, you old rascal; now I want to know what you've done with the ham that was in it?"

"Ham. Wacher talkin' bout, chile? I don't know know nothin' bout no ham."

"You *don't,* don't you? Do you see this bull whip?"

"Yessah, I sees it."

"Well, if you don't get that ham and get it mighty quick, I am going to wear it out on your black back."

"Now, Marse Jim, yo' knows yo' ain't never gwine tech yo' po' ole Uncle Ike with dat whip for sumpin' dat he never done, jest case Tige find a meal sack behine my chimbly. Fore de Laud dey ain't no stan' fo' eny sech talk as dat f'om no black nigger, pahson or no pahson, I ain't stole no ham an' if dat's all yo' come fo', yo' jes as well go back where yo' come f'om an' lemme alone," snapped Uncle Ike.

"Now, don't yo' go to gitten mad, deacon, I wasn't 'cusin' you 'bout stealin' de ham, but dat what's deys got yo' in here for, ain't it? I was jes' goin' to tell yo' dat ever sence you bin here in de jail, Lisa and me is been prayin' dat you would hab faith and trust in de Lord to help yo' bear up under dis trial, and git yo' outen here."

"Now I'se gwine to tell yo' right dis minit, pahson, dat de *trial* ain't bother'n' me a bit, it's bein' locked up in dis stinkin' jail 'thout nothin' to eat but some mis'able scraps dat looks an' smells lack dey's bin tooken out o' de hog trough, an' look at dis pile of rotten corn shucks whar I got to sleep wid nothin' but a nasty ole hoss blanket, full o' sweat an' hoss hair to kiver myse'f wid and what wid de cold and de stench an' de fleas, I don't git no sleep a' tall but jes' tumbles about an' scratches all night long. An' now you comes along, preachin' to me 'bout havin' faith an' trustin' de Lord to git me outen heah. Ef you'd brung along a pone o' cracklin' bread an' a dram or two outen dat jug in de chimbly corner, it would o' done me mo' good dan dat prayin' dat you says yo' bin doin', but yo' ain't dat kind of nigger. Yo' jes' takes all yo' can git, an' don't give nothin' back 'cept a mess o' dang fool talk dat don't count. So yo' kin jes' git out o' dis an' you need'n' kim back."

"Well, deacon," said the Rev. Jenkins, "ef dat's de way yo' feels about it, I reckon I'se jes' be gittin' along, but I was gwine to tell yo' dat, eber sense yo' bin locked up in de jail, Marse Jim is had a half-dozen of de young niggers wid dat dog Tige huntin' all over the kentry fo' de hidin' place o' dat ham, an' yo' knows yo'sef, what a nose dat houn's got. He's sho goin' to smell out dat ham fo' he quits, den what good de ham goin' do yo', huh! Now I was gwine to say dat, ef yo' *spicioned* whar de ham is hid, an' let me know, I'd go git it and take keer of it, till you's gets out o' heah, but since I bin gittin' nothin' but 'buse f'om yo' I'll jes' git along 'bout my own bisness."

With these parting words, the parson started for the jail door. He had raised his cane to rap for the jailer to open it, when he was hailed by Uncle Ike with, "What yo' goin' off lack dat for, pahson. You needn't be in sich a dang big hurry. Come back, dey's sumpin' I forgits to tell yo."

"Oh! dey is, is dey?" said the preacher. "Well, git along and tell it, but I ain't goin' to listen to eny more talk like I'se been gittin' f'om you to-night, ham hid aroun' dis house; ef dey was, you knows yosef dat dog's nose mitey soon smell it out. Tell yo, Marse Jim, some o' dem no count, crap shootin' young niggers is done stole dat ham and throw'd de sack roun' behind de chimbly dar to make yo' lay de blame on me, so dat jest settles it fo' a fac'."

"That would settle it, Uncle Ike, you sly old nigger, if it wasn't for that big grease spot on your shirt, right between your shoulders. That ham came from my smoke house on your back, so you had just as well produce it."

'Why, de Laud bless yo', Marse Jim, dat grease ain't no ham fat."

"What is it then, you old whelp? Where did it come from and how did it get on your back?"

"Well, sah, it was jes' dis way, las' night while I was down on my knees longside dat cheer — lack I is every night, wraslin' wid de Laud in prar, dat grease just dripped down 'tween my shoulders from dat side o' cat-fish bacon yo' see hangin' dar f'om de rafter. Dat's how come it, sah, fo' sho."

"By golly, Uncle Ike," laughingly said his captor, "you would have made a darn good lawyer, but it won't do, old man, so if you won't bring in that ham or tell me where it is, I'll just take you over to

Judge Alston and see what effect a week or two in jail will have. March."

When the good old darky was taken into court and tried upon a charge of robbery, he was found guilty and in spite of his protestations of innocence, his tears, and prayers for mercy, sentenced to jail until he was willing to reveal the hiding place of the stolen ham. The jail was a dismal, ill-smelling little brick building with a dirt floor, with only one small window, just under the eaves on each side, and a drain ditch under the wall at one end; consequently light and air were almost negligible quantities. Uncle Ike had been locked up here for two days, and was lying on a dirty shuck pallet in a corner of his dungeon, weeping, praying and bemoaning his hard luck. He had about decided to give up and divulge his secret to the jailer, when he heard the door creaking on its rusty hinges, then the sound of a familiar voice came to him from out the darkness.

"Whar you is, deacon?"

"Sat you, pahson? Whar yo' reckon I is, over to my cabin, cookin' yo' a chicken suppah, huh? Wacher doin' heah, nohow?"

"Well, Br'er Isik, ever since yo' bin in de jail, count o' dat ham yo' stole, me an' Lisa is —"

"Now look heah, Pahson Jimkins, I ain't goin'

it sounds like I'se de one dat stole de ham stid o' yo'. Now, what you forgit?"

"Well, it was dis way, pahson. De night de ham was tookin, I was settin' by de fiah readin' in de Good Book 'bout de way Lisha made de two b'ars eat up dem forty, sassy chillun, case dey was gittin' in his way an' callin' him ole bald head."

"Never mine 'bout dat, Isik," snapped Rev. Jenkins. "What you forgit?"

"Well, den dey comes a mitey strong smell o' brimstone thru de winder an' when I gits up to shet de winder down to keep out dat smell, I saw dat Marse Jim's smoke house was lit up jes' as bright as day. I knowed Marse Jim wasn't in dar at dat time o' night, so I grabbed my stick and dodgin' roun' de trees and tru de bresh, I sneaked over to de sasfras bushes 'long de creek 'hind de smoke house. Den wid my stick all ready, I creeps roun' to de do' 'spectin' to fin' one of de niggers stealin' de meat, but when I looked in, 'stid of a nigger dar was de ole debil hisse'f pokin' a ham an' a string of sasages in a meal sack, so I —"

"Wus dey sasages, too, deacon, sasages?" interrupted the parson.

"Don't I jes' tell yo' dey was sasages! Shet up an' lemme go on. So I jes' drapped down on my han's an' knees an' crawled back 'mong de bushes. I ain't been hidin' in dar more dan 'bout a minit when he comes out wid de meal sack on his shoulder an' starts wadin' up de creek. When he gits to whar de holler end o' dat ole poplar log is stickin' out over de water he tuk de ham and de sasages outin' de sack, an rams 'em in de log, den he turns roun' an goes back de way he kim."

"What happen den, Isik, what happen den?"

"Why, den I wakes up settin' in dat cheer shakin' like I'se got de ager wid de sweat porin' offin' me lack I'se bin wadin' in de creek myse'f. It was jes' a dream I had, pahson, an' dat's all."

"*No, sah,*" said the parson, slapping Uncle Ike on the back. "No, sah, dat wusent no *dream,* deacon, dat was a vishun yo' had, a sho nuff *vishun.* Dat's what! Now, I'll jes' go long and git de ham an' de sasages fo' de moon gits up and when I lays my han's on 'em, de debil hissef ain't goin' fine 'em. You jes' keep a prayin' an' a shoutin', Isik, and 'twon't be long fo' dey turns you loose, but don't yo' tell anybody 'bout dat dream." Then the parson went on his way, singing at the top of his voice,

> "Dar's mighty good pickin's
> In de ole Holler log,
> And dey's gwine be nuffin' **left**
> Fo' dat mangy ole hog."

"So dat's de way de crow's flyin'," soliloquized Uncle Ike. "Dis nigger ain' gwine git none of de pickin's, huh, mebbe so, mebbe so, but ef he don't, dat ole pole cat ain' gwine waller *dem roun' in his black mouf so dey'll do him much good.*"

"How you fine yo'sef dis evenin', deacon," greeted the parson the next night. "I hopes you's feelin' better dan you was yistidy."

"No, I'se jes' 'bout de same, pahson, no better, no wusser. I reckons yo' gits de ham an' de sasages all right, huh?"

"Dat's what I drapped in to tell yo' 'bout, Isik. Dat wusent nothin' but a dream yo' had arterall, dar was nothin' in de ole log 'tall."

Now Uncle Ike was a wise old darky, and knew better than to arouse a suspicion in the mind of his pastor by accusing him of lying or in any way casting a doubt upon the truth of his statement.

"Well, pahson," said he, "it cain't be helped, and I recon we'll hab to git along widout 'em, but you 'member I tole you it was jes a dream an' now you better be gitten along case I wants to get some sleep."

When the jailer brought in his supper Uncle Ike said to him, "Sheriff, I wish you'd go and bring

Marse Jim over here to de jail, I wants to tell him whar dat ham is went to."

"Oh, yo' do, do you? Well, tell me and it will be the same as telling him."

"No, suh, dat won't do a'tall, Sheriff. I'se got to tell hissef."

"All right, Ike. I'll bring him over in the morning," said the jailer.

"Dat won' do needer, suh, case den it may be too late, so please, suh, go git him to-night."

When Jim Maples entered the jail, he hailed the old darky with, "Well, Ike, you have concluded to tell me where you've hid the ham, eh?"

"No, Marse Jim, I ain't gwine tell you no sich thing, case I never *took* no ham, but I is gwine to tell you whar it's went. It was dat black, frog mouth Pahson Jinkins what is stole de ham, suh."

"What, the *preacher?*"

"Yas, suh, dat's what I is tellin' you. He is de one."

"Then why didn't you tell me before, Uncle Ike?"

"Well, it's jes dis way, Marse Jim, ev'ry nite since I bin locked up he comes to de jail, and after a lot o' fool talk about prayin' de Lord to help me outen here, advises me to hab faith an' not let on to nobody dat I knows anything 'bout who stole de ham; axes me a whole mess o' questions about it hissef, wants to know where Tige fine de meal sack and what yo' done wid it, and if yo' missed any o' de sasages as well as de ham. Dat talk 'bout prayin' me out o' jail don't count for nothin', fur I knows de Lord ain't gwine take up His time wid no sich foolishness, but when he axes me whar Tige fines de meal sack an' what you done wid it, an' ef you misses any o' de sasages, I 'spicions dat nigger knows just whar dat ham is went, but my 'spicions ain't no proof, so I jes' lays low an' holds my tongue till I gits dat proof, an' now I'se got it, Marse Jim, an' dis is it." He then handed Jim Maples a pone of corn bread and said, "He gimme dat cracklin' pone jes befo' he left to-night, an' you can see it is soaked plum full o' sasage gravy an' dey's a little piece of ham in de middle of it. Now, whar did he git the sasage to make dat gravy, an' whar de ham come from? Why, of course he stole 'em outen your smoke house. Now, Marse Jim, it's along about time o' nite dat de pahson an' Aunt Lisa has dey supper an' as dis is de Quarterly Meetin' ob de Church an' de Persidin' Elder is stayin' dar, an' Aunt Lisa will git him de best supper she knows how to cook. Dey'll have everything good she can skeer up, 'sides de ham an' de sasages, an' a jug o' cider out o' de bar'l down by de corn crib, to top off wid. So ef we goes over dar now, we's gwine to ketch 'em right in de middle ob de feast, fo' when dem three niggers sets down to dat table an' gits to eatin' an' drinkin' dey ain't goin' to be in no hurry to quit, as long as dey is anythin' good to swaller."

"By golly, Uncle Ike," said Jim, "you are some detective sure, and I guess you have spotted the thief all right. Come on and let's get over there and nab the old rascal."

The parson was in the act of helping his guest to a thick slice of juicy ham and Aunt Lisa was refilling his cup with cider from the jug when the raiding party entered the room.

"So, you black limb of Satan," rasped Jim, jerking the parson to his feet. "You are the thief who has been robbing my smoke house and draining my cider barrels."

Then turning to the elder, he said, "Elder Cook, I guess you will have to get another pastor for the church here, this black rascal will have a long vacation in the jail, in which to pay for his stealings." At this junction, the parson wailed out, "Yo' is wronging me, Mister Maples, I'se never stole nothin' from you or nobody else."

"Oh, is that so? Then where did you get the ham and the sausages?"

"I fine dem, sar, whar dat no count nigger Ike hide 'em," said the parson.

"Oh, you *found* them, did you? If that is the case, why didn't you tell me that you had found them, instead of appropriating them to your own uses?"

"How I know dey is de ones you is lost," mumbled the preacher.

"Well, we'll let it go at that for the present, and go on over to the jail where yo' will have ample time to refresh your memory."

When Jim and the jailer started for the door with their prisoner, Uncle Ike left the corner, where he had stood grinning and slapping his leg gleefully at the parson's predicament, touched him on the shoulder and said, "Dis ain't no dream, pahson, it's a revelation, dat's what."

They Broke the Jug and Quit

BILL GUTHRIE and Tule Jim had been friends for many years. They lived together, hunted and fished together and got drunk together. They were wood choppers and had, at the time of this writing, finished a contract for cutting one hundred cords and had received their money therefor.

"Say, Guthrie, was you ever at the State Fair down to Sacramento?"

"No, I wasn't; what you want to know for?"

"Well, I been readin' in the paper here that there's going to be some mighty big doin's goin' on down there when it opens in September. Hoss racin', cow punchin' an' all sorts of games. I never been there, myself, and I was just a-thinkin' that by the time we cuts another hundred cords o' wood, everything will be in full swing. By that time we'll have money enough to go down and take it all in. We can dress up in some good clothes, get us a watch apiece an' jine the procession without axin' no odds from anybody. Why, Bill, it's the biggest chanst we ever had for seeing things and having a good time. If the idea strikes you all right, I reckon we better give what money we got to the boss to take care of for us till we gits ready to go. What do you think of it?"

"I don't think nothin' 'tall of it, and I'd just like to know how you got that crazy notion in your fool head. Don't you know, you danged ole Tule Poop, that 'bout all the pickpockets an' tinhorn gamblers in the state will be down there while the Fair's running? I tell you, Jim, if you git mixed up 'mong that crowd of sharps with marked cyards an' their loaded dice, you'll run into the biggest snag you ever struck in your life, and the fust time you set down in a poker game you'll be soaked up with doped whisky and then skinned slicker than a rotten onion. Now if this is your idea of a big time, you can jest go it alone, I don't want any of it.

"What's the matter with Sharp's Point? It 'pears to me we's always had a good enough time with the boys up there. We knows 'em all and they's all our friends. They's never any marked cyards rung in on us in a poker game and if we loses our money we loses it on the square, and we loses it to our friends and we's got a chanst to get it back in the next game if we hold the cyards. We can drink all the whisky we wants without gittin' pisened, too. For the brand Jim Gurley hands out is the pure stuff and won't hurt nobody."

"Well, Bill, we's always been pardners, an' we's pardners still, an' if you says Sharp's P'int, Sharp's P'int goes with me. When you reckon we better go?"

"Lemme see, the races begins on Thursday, an' this is Monday. The boys will all be gittin' together by now. I'm figerin' we jest might as well go on up in the mornin' so's to shake hands all roun', take a drink or so with maybe a game or two o' poker before the races begins."

After playing the races, losing the most of their money to their friends at poker and soaking themselves in Jim Gurley's "pure brand" of whisky, they staggered up to the boss the next Sunday night, carrying a jug of whisky Jim Gurley had given them when he told them good-by, and greeted him with, "Hello, boosh. Howsh thingsh? You see we'sh got back."

"Yes, I see you have," said the boss. "And you have got back with a full cargo of 'bug juice' as usual. How much of your money have you brought back with you?"

"Why, you shee, boss, Jim Gurley's ole wooden

THEY BROKE THE JUG AND QUIT

leg ish about wore out, and we give him the money we'sh got left to hep him git a new one. Jimsh mighty good fren; he'sh give us this jug o' whisky for nothin' a' tall. Take a drink, boss."

"No, thank you. I don't want any to-night, nor do either of you. You need sleep more than you do whisky, so put your jug over there in the chimney and go to bed."

"All right, boss. What you say goes. Goo' night." With these parting words, the men spread their pallet on the floor and were soon in the land of dreams. When the boss heard them snoring he took a pitch fork, going to a spot some distance away, where on the day before he had found a dead pig in the last stages of decomposition. Impaling it on the fork, he carried it to the house and shoved it under the floor to within a foot of the heads of the sleeping men. He then went quietly back to his room and waited for what would happen. He had been listening for, perhaps, ten minutes, when,

"Whew," came from the other room. "God'll mighty, what a breath. Guthrie! Say, Guthrie! You danged old hog, for God's sake, turn over. Lord, man, you're rotten."

"You're a liar, you dirty houn'. You smell yo'self. I been layin' here for more'n an hour, with my head covered up an' holdin' my nose 'count of the stink that's been comin' from you."

"You're a liar yo'sef, you ole polecat. You smells lack you's loaded up on a lot of rotten aigs an' if I had a breath like you I'd go jump in the river an' drown mysef."

"I've took all I'm gona take from you, you dang'd 'Fly up the creek.'" With these words, Guthrie struck out and landed a stiff punch on Jim's nose. Jim came back with a hard jab to the stomach, causing Guthrie to lose his breath and double up with pain. Then they clinched and began fighting all over the room. When the fighting began the boss entered the room and shouted:

"Here, stop that. I won't have any fighting in here. There is no occasion for it, anyhow. I have been listening to your quarrel and have heard all the vile epithets you have applied to each other. You have both told the truth about the other's breath, and I tell you boys your condition is a very serious one. Your stomachs have already begun to putrefy and unless you quit your whisky drinking you will soon be dead men."

"My God, boss," said Tule Jim. "You don't mean to say it's as bad as that, do you?"

"Yes, it is, and I am not sure if quitting the

whisky now will save you, but it is your only chance, and I advise you to do so right now."

"Guthrie," said Jim, in a very shaky voice. "Le's b—bust that dang'd jug."

"All right, Jim," answered Guthrie. "We'll d—do it." And busted it was.

The next morning the boss took the pig from under the house and buried it.

Bill and Jim never knew that it was the smell of the pig and not their bad breath that caused them to become "Sons of Temperance."

A Sudden House Moving

HE was a little Holland Dutchman. He had been a swine herder in his native land and when he came to California in 1850 naturally sought a job at the same work. He found one with Colonel Robert Beasley and my father, who owned the Twin House Ranch on the Sacramento River. In 1851 my father bought all the hogs and shipped them to San Francisco, where he had them slaughtered and sold in the markets of that city, Henry, of course, going along to care for them. Now Henry had a great and abiding love for his herd, collectively and individually, and it nearly broke his heart to see them butchered. He would never help in this work, but would go off to some place out of hearing and mourn over the loss of his beloved pigs. When the herd of four thousand had been reduced to something like one hundred, Henry located forty acres of land on the Potrero, bought the remaining hundred and started a hog ranch of his own. His house was built on a steep hillside, the floor being some three feet above the ground on the lower side and one foot on the upper, boarded all round with the exception of a hole three feet wide, facing down hill. His object in building this way was to provide comfortable sleeping quarters for his hogs. When he got settled in his new house with his beloved porkers around him, he was supremely happy and would, in the evenings, sit and talk to them and sing songs of the "Vaterland" for them until bedtime.

As long as the weather kept warm, the hogs behaved themselves as well as hogs could be expected to, many of them preferring to sleep in the open air, but when winter came on they would all crowd under the house.

One bitter cold night with the wind blowing almost a gale and the rain coming down in torrents, they were under there crowding, fighting and jostling each other and squealing to "beat the band," every one trying to get in the middle. Poor Henry became almost beside himself at their awful racket and began pounding on the floor and shouting, "Schtop it, you son of a guns, schtop it. I tole you. Does you want to knock down der house, you tam hogs."

His blows on the floor and his commands to quit had not the slightest effect on the hogs. They kept right along until at last Henry became so enraged that he built a fire, and filling a wash boiler with water, placed it on the stove, and putting his fingers in his ears, sat down on the side of his bed to wait for it to get hot. When the water got to boiling he lifted it from the stove and set it on the floor, and then, once more, shouted to the hogs to stop their noise, with the same result as before.

"Vell, den, if you vill not do like I tell you vonce, I vill gif you someding to squeal mit, vot!" So saying, he turned the boiling water over the floor. Then pandemonium broke loose. Woof! Woof! came from under the house; then, with one concentrated squeal the hogs broke for the exit, so many of them reaching it at the same time that the underpinning gave way and the lower side dropped to the ground, closing the hole, trapping most of the hogs on the under side. Then the house was lifted bodily from the ground and went swaying and cracking down the hill, with Henry inside, buried under his household goods, having the ride of his life.

When the house finally stopped, it had been carried on the backs of the hogs full two hundred yards down the hill, and entirely beyond Henry's boundary. It had landed in an offshoot from Mission Creek and when Henry scrambled out from the wreckage he saw the hogs wallowing and rolling over and over in the mud, still grunting and squealing from the pain of the scalding water.

"Now, look vot you haf done vonce, tam you. I haf safe you from der butcher, I haf gif you blenty to eat, I haf make you a nice varm place to schleep mit und blanted scratchen posts all ofer der ranch vere you can scratch dem fleas off mit, und now, you pay me back by pull mine house by der hill down und break up everyting vot vas by der inside. After dis you vill schleep out in der cold und der rain und by chiminy you gets no more dot house under. I am done mit you, tam you."

How Uncle Bob's Prayer Brought the Rain

"WHY, how are you, Uncle Joe? I am sure that I am glad to see you again. I didn't know what had become of you. Where have you been keeping yourself for so long?"

"Ise bin reslin wid de rumatis, fur a long spell now, sah, an' I ain't bin able to git out no whar. Ef it had'n bin fur dat, sah, I bin come to see you long fo' dis."

"I am very sorry to hear that you have been so afflicted, Uncle Joe, and had I known it, I would most certainly have been to see you. Have you been doing anything for your rheumatism?"

"Yes, sah, I'se bin rubbin' my back an' my laigs wid sum hoss medicine dat de livry stable man gin me, but it don' seem to hardly do me no good, but jes burns, lack fiah, whar I rubs it on; an' den big blisters cums along, an' dey makes me feel mitey so', sah; an' when dem blisters goes along I gits to itchin' all ober an' I has to keep scratchin' all de time jes lack I'se full ob fleas, sah."

"Well, you are feeling better now, Uncle Joe, aren't you?"

"Yas, sah, I'se feelin' tolerble peart dis mornin', but I cain't tell how long dat's gwine to las'. Sometimes I t'ink I'se gwine git aroun' again all rite, an' den dem pains comes along agin, an' I goes down on de flat ob my back an' dar I is. It's mitey bad, sah, mitey bad; but den yo' see, sah, I'se gittin' to be a mitey ole man an' I specs dat de Lawd's gwine to call me home fo' long, an' I'll jes hab to b'ar de pains an' git along de bes I kin, till dat time cums."

"How old are you, Uncle Joe?"

"I don't no jes zacly how ole I is, sah, but I nos dat it's ben a mitey long time ago dat I was born'd. I 'members dat me and Lira had just got mar'ed de time dat de stars all fell, an' I was som'whar long 'bout twenty den; so yo' see, sah, I mus' be gittin' long close to a hundred."

"You have lived to a ripe old age, Uncle Joe, and many strange and wonderful happenings must have fallen under your observation. I would like very much to have you tell me some of them."

"I don' no zacly what de obsvashun is, sah, but I'se seed lots ob t'ings in my time dat dey ain' no countin' for. Now dar wus dem stars fallin'! Dat beat anyt'ing I eber seed or hear'n tell about. It wus Miss Becky's berfday an' de young folks fum all ober de kentry wus habin' a dance in de big house; dat night Uncle Rance wus playin' 'Tuckey in de Straw' fur 'em an' dey wus for'd and backin', an' all hands roundin' an' also leftin', an' all promenadin' an' joyin' deysefs de bes' in de worl', when dey cum de orfullest screech fum outsid'n de house dat eber was hyeard tell on; an' twant a minit arter dat befo' Ant Susan cum a-tearin' inter de room yellin' jes' as loud as she could, 'Quit yo' dancin', chillun, quit yo' dancin'! De Jedgment Day is done come! Quit yo' dancin' an' git down on yo' knees an' go to prayin'!"

"I tell yo', sah, dat broke up de dancin' mitey quick, an' dey all reshed to de do' to fine out what's de matter. When dey go dar dey seed de mos awfules' site dey eber laid dey eyes on. Dar wus dem stars droppin' down outen de sky by de milyuns; big ones an' little an' middle-sized ones; shootin' dis way and dat way, an' a bustin' an' poppin' jes lack dey's goin' to 'stroy de hole worl'."

"Were the people frightened, Uncle Joe?"

"Does you mean wus dey skeered, sah? Dat wus de wust skeer'd lot ob fokes dat was eber seed in dem parts; de niggers all clim unner de house an' hid deysefs dar, an' de white fokes ack lack dey done gone plum' crazy. Some ob dem wus a shoutin' an' a-prayin', som ob dem wus grabbin' holt on an' pullin' one anoder round de yahd, an' som ob de oders wus layin' flat on de groun' wid dey faces in de dirt moanin' an' cryin' lack dey los' eberting dey had in dis worl'. An' African Jim, de blacksmif, cum bilin' outen de smoke house wid a big ham he bin stealin' in dar, an' wen he seed dem stars fallin' all roun' him, he begin dodgin' dis way an' dat way lack dey was sumthin' tryin' to ketch 'im, all de time hollerin' out,

"'Oh, Lawdy, oh, Lawdy, de debil's gwine to ketch me! Don' let 'im take me, Mars Jeems, don' let 'im take me an' I'll jes be de bes' nigger arter dis, dat eber wus, an' I'll neber steal no mo' ob de chickins no' de bakin'!'

"Den he drap dat ham an' broke fur a big sugar hogsit dat wus dar in de yahd an' dive hed fust in de hole dat wus in de side; an' wen he try to turn de hogsit so de hole wud cum nex' to de groun', dat hogsit start rollin' down de hill an' it kep' goin' faster till it wus goin' roun' lack a spinnin' wheel. Jim bin rollin' roun' so fast an' gittin' sich a bumpin' while he's in dat hogsit dat wen it run against a big stump an' stop, wid de whole side busted in, he crawled out lookin' lack he don' no whar he's bin o' what he's bin doin'. He wus gaumed all ober wid de sugar dats bin stickin' to de sides ob de hogsit an' his wool wus smeered down on his haid es flat as a hoecake; I tell yo', sah, he wus a mitey po' lookin' nigger es he stan' dar wid his mouf wide open an' his eyes rollin' roun', trimlin' an' shakin' lack he's mos' skeer'd to death, an' I tells yo' dat he did'n lose no time gittin' under de house wen he kims to hissef. Yo' orter seed dat Jim, sah, when he crawled fum under dar; he wus jes kivered all ober wid chickin feders count ob de sugar dat's stickin' on his close an' his haid. Yo' could'n see notin' but de white ob his eyes an' he look mo' lack a nostich dan he does lack a nigger, sah. All de fokes was feelin' mitey bad count ob de stars all drapin' outen de sky de night befo' an' say dey reckons de moon goin' kim down nex', an' den de nites wus goin' to be turrible dark wid de moon an' de stars all gone, but when nite kim along again dar wus all dem stars back up dar in de sky shinin' es bright es eber, jes lack dey had'n drapped out de night befo'; an' how dey gits dar, dout nobody seein' dem jes' beats me, dat's sho'. Dat wus de mos' turablest site dese ole eyes eber looked upon, sah."

"Were you scared, Uncle Joe?"

"I don' no ef I wus zacly skeer'd, sah, but I wus mitey 'cited, least ways, I wus under de house 'mong de oder niggers."

"That must have been a magnificent and awful sight, Uncle Joe, and I don't wonder that you were excited."

"Now, Uncle Joe, I want to have a little talk with you on a present-day topic. What, in your opinion, has been the cause of so little rain for the past three years?"

"Well, sah, de mos' I kin make outen it es dat de people hab done gone away fum de Lawd, an' insted ob goin' to chuch an' prayin' Him to sen' de rain, ever las' one ob dem hab bin tryin' to grab all de money dey kin lay dey hans on, an' dey don' keer ef dey gits it hones' or no, so dey gits it, is all dey wants. Jes' look what dey's axin' fur vittles, sah; datle tell yo' how greedy dey is fur de money. Fur one t'ing, dar's de hog meet; fo' de wah I could git de bes' kine ob good, fat bakin' fo' a bit a poun', but de way dey's sellin' it now I cain't git nuf bakin to grease de fryin' pan, let lone eatin' it. 'Tain't case de hawgs is scase, sah. Dey jes' loves de money better den dey does dey own souls, an' dey's gwine to hab it no matter what cums ob de po' fokes. I jes' lack to no, sah, what dey's gwine do wid all dat money when dey gits it. I noes dey is habin' a big tearin' all ober de kentry wid dey ottermobeels an' sich lack, but dat ain' goin' last foeber. Dey's gwine die sum ob dese days, an' dey money won' do dem a bit o' good in de nex' worl', dat's sho'."

"All that you have said regarding mankind's mad race for money is true, but do you believe that God would send the rain if the people prayed for it?"

"Yas, sah, I sholy does."

"Do you know of an instance when a prayer was answered for rain? If you do, tell me about it."

"Well, sah, dar wus de time ob de big drouf down dar in Georgy, when Uncle Bob Alston gits de bigges' answer to his prar fo' de rain dat enybody eber hearn about befo'."

"Who was Uncle Bob Alston, Uncle Joe?"

"He wus an' ole nigger preecher, sah, an' bilonged to Marse Wilyum Alston. When Marse Wilyum went ober to de colige at Macon to learn to be a preecher, ob cose he took Uncle Bob 'long wid him to bresh his close, blak his boots an' eber ting else dat he could to make Marse Wilyum cumftable. Dey wus dar at de colige fo' years, an' all dat time Marse Wilyum bin readin' de Bible to Uncle Bob ebry nite, so dat by de time dey gits ready to go back home he knowed mos' as much 'bout de insides ob de good Book as Marse Wilyum did hissef. Well, sah, when dey gits back home from de colige Marse Wilyum wus pinted to preech all ober dat part ob de kentry; so him an' Uncle Bob, dey started out to do de Lawd's work, Marse Wilyum preechin' to white fokes an' Uncle Bob to de niggers. When dey bin preechin' like dat fo' long 'bout five yeers, dey wus another big meetin' ober dar in Macon of all de Methdis preechers fom almos' ebry whar, an' case he knowed mo' bout preechin' dan all des res' ob dem put togeder, dey make him de Bishup ob de hole lot. De year arter dat Marse Wilyum tended de big Confrace up dar in Nashville whar mos' all de Methdis preechers in de Nited States wus got togeder to talk 'bout de chuch bisness, an' cose Uncle Bob went wid him. Dey had'n mo' an' got good away, sah, when de people begin to act jes' lack dey did'n owe de Lawd nothin' 'tall. Insted ob goin' to meetin' and habin' de prar meetin's an' de love feests an' sich lack, dey jes' broke loose an' gin playin' cyards an' racin' horses an' drinkin' whisky at de tavuns, an' havin' dey dances an' frolics on Sunday nites; dem wus de white fokes, sah. De niggers wus breakin' into de smoke houses an' de corn cribs an' stealin' all de chickins dey could lay dey hans on sides goin' fishin' an' possum huntin' on Sunday."

"The negroes had quite a reputation for stealing chickens, didn't they, Uncle Joe?"

"Yas, sah, dat's so, but den dem niggers did'n hab no chance to larn notin', sah; dey wus jes' lack little chilun, an' ef dey cum across a nice fat rooster settin' on de pole neerest de groun' it look to dem lack he wus jes' put dar a pupose so dey could kech him easy; an' dey say to deysef, ef dey don' take him ole Mr. Coon gwine git him, so dey jes' grabs him an' takes him home. Chickins is mitey good eatin', sah, ef dey's fixed up rite. Yo' jes' take a nice fat rooster and fill 'im up wid good cohn bred stuffin', wid a plenty inguns, some sage an' pasley an' a sprinklin' ob black pepper in it, an' den put him in de Dutch oben wid sweet taters all roun' him an' roas' him till he's nice an' brown an' jes' swimmin' in de gravey, I tell yo', sah, dat's jes' de bes' eatin' in de worl' dat I noes on, 'cep' de possum."

"Did you ever take a chicken off the roost to keep a coon from getting him, Uncle Joe?"

"I'se done tole you, sah, how Lira fixed dem up, so dey's mighty good eatin', but dat wus sich a long time ago dat I done forgit how I kim by dem.

"Well, sah, when de wicked doin's I ben tellin'

yo' 'bout bin goin' on 'bout two weeks, de craps begin showin' dey wus need'n water. De leabs ob de cohn wus wroppin' deysefs roun' de stalks, an' de cotton bolls wus droppin' close to de groun'; de leaves wus drappin' off'n de trees an' de grass wus lookin' mitey po'. Den de water gin to dry up in de creeks an' de pon's. By dis time, sah, de fokes begin gittin' mitey oneasy, but dey don' min' dey ways, but kep' rite long wid dey wickedness, an' case dey was gittin' scouraged dey wus drinkin' mo' whisky dan eber. At las' de pon's wus all dry an' dey wusn't a drap runnin' in de creeks an' branches cep de Big Creek, an' dat had jes a little trickle creepin' down de middle; ob cose, dey wus some holes in de creeks whar dey wus considable water, an' de fish in dem holes wus jes' 'bout keepin' alive an' dat wus all; dar dey wus swimmin' close up to de top wid dey noses stickin' outen de water, workin' dey moufs an' a-pantin' lack dey wus wantin' a drink mitey bad. De mud turkles wus all borin' down in de bottom of de pon's tryin' to fine some sof mud whar dey could cool deysefs off, an' eben de ole bull frogs wus so thirsty, sah, dat dey couldn't beller, an' jes' sot dar on de logs blinkin' dey eyes at de sky an' lookin' mitey sorry. De niggers, dey bin killin' de black snakes an' hangin' dem in de hazelnut bushes all ober de woods tryin' to make de rain come dat way, but it ain't no use; an' it kep gittin' dryer an' hotter all de time, an' everybody wus feelin' mitey bad, case it look lack everting goin' to be plum stroyed wid de heat an' de drouf. When tings gits in dis way, sah, Marse Wilyum an' Uncle Bob dey comes home, an' when dey seed de dishun de kentry wus in, count ob de way de people hab lef' de Lawd an' bin follerin' arter de debil whiles dey wus away, dey bofe looks lack dey hearts wus broke fur good. De nex' mornin', Marse Wilyum sen' five o' six ob de yung niggers out ober de kentry to tell ever'body to git up to de meetin' house de nex' Sunday mornin', case Uncle Bob was goin' to pray de Lawd to sen' de rain, an' when Sunday mornin' come along de hole kentry wus jes' kivered wid de people gittin' to de meetin' house; de white fokes in cyarages an' on dey hosses, an' de niggers a walkin'; an' dey wus so sho dat de Lawd ud answer Uncle Bob's prar dat ebry las' one of dem brung along dey umbrelers wid dem. Well, sah, when de meetin' wus opened de chuch wus cram plum full ob white fokes an' dey wus hunduds ob niggers on de outside. Arter de congergashun singed a song Marse Wilyum an' Uncle Bob goes 'long up in de pulpit, an' gittin' down on dey nees dey say a prar apeece; den Marse Wilyum open de Bible an' read a chapter outen it. Arter dat he begin talkin', an' de way he bused dat congergashun fur dey cyard playin' an' de whisky drinkin' an' dey hoss racin' an' dancin' on Sunday; an' dey goin's away fum de Lawd and takin' up wid de debil finally, he make dem fokes dar in de meetin' house feel lack dey wus de wust lot ob trash on de face of de yearth. Den he go on an' tell dem how mad de Lawd wus wid dem fur dey wickedness; an' he tell dem dat he don' 'no' ef he gwine forgib dem or no', but dat Uncle Bob was gwine to pray de Lawd to sen' de blessid rain to let dem 'no' ef he's gwine let dem off dis time fo' dey sins. Den dey all gits down on dey nees an' Uncle Bob begin talkin' to de Lawd jes' lack he wus rite dar wid dem in de chuch. He tole Him 'bout all dey carins on an' dey cuttin' up while him and Marse Wilyum wus up dar in Nashville.

"'Dey wusn't doin' nuthin', Lawd, while we's tendin' yo' bisness but jes' doin' eberting dat de debil tole dem to do. Dey wus breakin' in de smoke houses an' de cohn cribs, stealin' de cohn an' de bakin an' robbin' de hen houses ob ebry chickin dey could lay dey han's on, an' den, Lawd, when ebryting begin dryin' up dey goes into de woods an' kills all de black snakes dey kin fine an' hangs dem up in de forks of de hazzlenut bushes to fetch de rain dat away. Deys bin turable wicked Lawd, an' I noes you is mitey mad 'bout it an' I don' blame yo' a single bit; an' et would sarve dem rite ef yo' wud sen' de debil to tote some ob de wust ob dem off wid 'im; but den ef yo' done dat, Lawd, de white fokes wud loose dey niggers an' dat w'u'dn't do a tall, so I is, umbly, axin' yo' to forgib dem jes' dis once fo' yo' noes yo'se'f dat dey is notin' but po' ignunt an' don' 'no' notin'. An' now, Lawd, Ise axin' yo' to sen' de blessed rain down fur we's needin' it mitey bad. I ain't axin' fur no gentle showers, Lawd, fur dem ain't what we wants. We needs de water, Lawd, we needs de water. Sen' it down by de bar'l full, in sheets an' streems, till de creeks an' pon's es runin' ober an' de yearth is soaked plum thru.'

"Well, sah, Uncle Bob prayed along lack dis fur 'bout a nour but it don' seem lack et goin' to bring de rain. Dey wus some ob dem no 'count snake killin' niggers setten up by de do'; when dey see t'ings lookin' lack dis dey begin talkin' 'mong deysefs 'bout Uncle Bob, an' makin' fun ob him. Efe Jinkins, he say to de oders,

"'Ise t'inking dat ole fool, Uncle Bob, ain' goin' bring no rain dis time wid dat fool prayin' he's doin' in dar.'

"I specs, sah, dat Uncle Bob hearn dat nigger,

when he say dis, case he jes' looked up lack an' say,

"'Lawd et ain't goin' do no harm ef yo' sen' de debil and let him take away dat no 'count, chickin stealin' nigger, Efe Jinkins, out dar by de do' an' put him down whar he belongs. He ain't wuth nothin' no how an' his master wud git shet of a mitey po' nigger.'

"When dey hear Uncle Bob talkin' to de Lawd dat way, dat lot ob niggers git away f'om de do' lack dey hear de oberseer blowin' de horn fur dem to go work in de cotton patch; an' yo' jes' orter seed dat Efe, sah. He wus a coal black nigger, but he wus skeer'd so bad dat he turned jes' de color ob ashes; an' his eyes dey pop out lack de eyes ob a big swamp rabbit; den he gits down on his all foes jes' as close to de groun' es he kin an' crawls roun' de cohner ob de meetin' house tell he gits out ob site; den he gits up an' brak's fo' de woods lack de debil wus reachin' fur his coat tails, an' hid hisse'f 'mong de underbrush; an' I tell you, sah, dat nigger didn' show hisse'f no mo' at dat meetin'.

"All dis time de sky wus jes' de color ob a brass kittle; dey wasn't a bref ob air stirrin' an' de sun was jes' as hot as fiah. Dey wus no sine of de rain comin' yet, but Uncle Bob kep' on prayin' lack he wus boun' to hole on to de Lawd tell He ans'er his prar; an'

I specs, sah, dat de Lawd wus jes' standin' dar lis'-enin' to Uncle Bob, tell he fine out how much rain Uncle Bob want, fo' he turn de water loose, case it wus only a little arter dis dat a little cloud, no bigger dan yo' han' riz up in de wes'; den a little whirl-win' come 'long raisin' de dust in de road an' twistin' de leaves roun' an' roun' in de woods. Den dat cloud begin gittin' bigger an' bigger, an' blacker an' blacker, tell in jes no time a tall de sun wus shet out an' it wus mos' as dark es nite; an' den de win' riz up an' come roarin' an' tearin' thru de woods, cuttin' de lims an' de tops offen de trees lack dey wus nothin' but a patch ob Jimson weeds. Den all ob a sudd'n a streak ob forked litenin' shoot out'n dat cloud, liten up de hole kentry; den dey cum a big clap o' thun-der soundin' lack a milyun ob big guns; den de thun-der an' liten kim along rite togeder blarin' an' roarin' an' crackin' lack dey wus gwine split de yearth wide open. All dis time Uncle Bob kep' a-prayin'.

"'Sen' de rain, Lawd, sen' de rain!'

"An' all ob a suddin dat cloud 'peared to bust wide open an' here cum de water porin' down jes' lack de bottom ob de riber dropped out. Den de nig-gers hist dey umbrelers, but dey ain't hardly got dem up fo' de win' turn dem 'rong side out, an' dey jes' drapped dem an' broke for de under side of de meetin' house, an' when dey git dar ebry las' one ob dem try to get in de middle jes' lack de hawgs does ob a cole nite. An' I tells yo', sah, de way de rain wus comin' down wus a caushun; it wus jes' a-fallin' in big chunks lack it wus tryin' to mash de meetin' house flat on to de groun'; an' de litenin' wus flashin' an' shootin' in eber which away, stakin' a tree here an' a tree dar, breakin' sum ob dem rite in two an' peelin' all de bark offen de other wuns, an' de thunder kep' a-roarin' an' crashin' rite 'long, an' de way it made dat meetin' house rock an' trimble was a site to look at. What wid de thunder an' litenin' an' de rain, de fokes in de chuch was most skeered to death, an' dey was shoutin' an' prayin' an' cryin' lack dey wus gun plum' crazy. All dis time, sah, Uncle Bob wus prayin':

"'Let de water come, Lawd! Let de water come!'

"An' de way de water kep' comin', sah, was neber seed in dis worl' befo'; couldn' a rained no harder dan dat, sah, de time Noah got de Lawd to sen' de flood an' droun ebry body in de worl' fo' dey wickedness. By dis time, sah, de pon's an' de brenches was plum' full, an' de water in de big creek was comin' down wid de awfles roar dat wus eber hearn, bringin' a big poplar log wid it, whirlin' roun' an' roun' in de current wid sticks an' all kines ob trash racin' 'long tryin' to ketch up wid it. Den de fokes git skeered dey wus goin' git drounded fo' sho', an' dey begin to holler:

"'Hole on dar Uncle Bob; Hole on! Dat's enuf. Let de Lawd go!'

"Uncle Bob 'pear'd lack he didn' hear dem an' jes' kep' prayin' fo' mo' water. Den, sah, de water in de big creek gin runnin' ober de bank an' it kim slidin' ober de groun' tell de hole lan was kiver'd wid it; an' when it gits 'mong dem niggers under de chuch dey tink dey's gone fo' good, sah, an' dey wus sich a scramble to git outen de hole whar gits in, dat dey lack to lif' de meetin' house rite off'n de groun'. Uncle Bob wusn't payin' no tenshun to what's goin' on roun' him, but jes kep' on prayin' de Lawd fo' mo' rain. When tings done come to dis pass, Marse Wilyum gits down from de pulpit an' goin' ober to Uncle Bob an' layin' his han' on his haid, says:

"'I tink we's got nuff fo' dis time, Bob; de Lawd hab sen' a woneful ans'er to yo' prar, so git up now an' we'll sing His praises fo' His goodness an' mercy, till de rain stops.'

"After dat, sah, de thunder an' litenin' dey stop an' de rain begin gitten liter an' liter, an' by de time dey sing 'On Jordan's Stormy Banks' an' 'Praise

God From Whom All Blessin's Flow,' it stop altogeder an' de sun shined ag'in lack it was 'fo' de rain, on'y it wan't so hot. When de peeple kim outen de chuch, ebryting look mitey freshen up an' de birds wus singin' lack dey goin' split dey troats; de bob white's wus a-callin' to dey mates, an' de cafs wus runnin' an' jumpin' wid dey tails swingin' over dey backs lack dey wus feelin' mitey good case de water cim to make de grass grow. Den, all at once a great big rainbow 'peard in de east, stretchin' f'om one end ob de yearth to de oder; an' when de peeple seed dat, dey begin shoutin' an' singin' de Lawd's praises, case it wus a token f'om Him dat He done forgib dem fo' dey sins. When Marse Wilyum git close to de big house wid his fokes, dar wus Ant Becky — dat was Uncle Bob's wife, sah — standin' in de do', an' when she seed how de fokes wus draggled wid de wet an' de mud, she git jes so mad es a hohnit, an' th'owin' up her han's she holler'd out:

"'Jes, look dar now, dat's jes' de way wid dat fool nigger Bob. He never noes when he got enuf. He's jes' kep' holdin' an' boderin' de Lawd tell ever'body's mos' drounded, an' dat's de reesin I didn' gwine to de meetin'.'

"'Yes, sah, I suttinly does beleeb dat de Lawd ans'ers prar fo' de rain. I sho' does, sah.'"

Mark Twain, Bret Harte, Steve Gillis and dear old Dan McQuill are dead. Of all that gay company at Virginia City I alone remain. I think I shall soon see that company again.

THE END